SHACKLETON
Leadership Lessons from Antarctica

Arthur Ainsberg

iUniverse, Inc.
New York   Bloomington

Copyright © 2010 by Arthur Ainsberg

All rights reserved. No part of this book may be used or reproduced by any means, graphic, electronic, or mechanical, including photocopying, recording, taping or by any information storage retrieval system without the written permission of the publisher except in the case of brief quotations embodied in critical articles and reviews.

iUniverse books may be ordered through booksellers or by contacting:

iUniverse
1663 Liberty Drive
Bloomington, IN 47403
www.iuniverse.com
1-800-Authors (1-800-288-4677)

Because of the dynamic nature of the Internet, any Web addresses or links contained in this book may have changed since publication and may no longer be valid. The views expressed in this work are solely those of the author and do not necessarily reflect the views of the publisher, and the publisher hereby disclaims any responsibility for them.

ISBN: 978-1-4502-1537-4 (sc)
ISBN: 978-1-4502-1538-1 (dj)
ISBN: 978-1-4502-1539-8 (ebook)

Printed in the United States of America

iUniverse rev. date: 03/29/2010

This book is dedicated to those extraordinary souls whose lives have been devoted to exploring the unknown.

CONTENTS

Foreword	ix
Introduction	xi
THE VOYAGE	1
THE LESSONS	17
The Fire Within: Feel the Purpose in Your Gut	19
Be inspired – If you aren't, neither will your team.	
A Captain Needs a First Mate: Choose a Powerhouse for Your Number Two	29
Don't underestimate the power of a second leader whose know-how rivals your own.	
Can You Sing? It's All About the Team	37
Like a jigsaw puzzle, every piece is important in a group effort.	
Camaraderie at 20 Below Zero: Creating an Optimal Work Environment	47
Productivity flourishes in a nurturing, constructive atmosphere.	
Sailing Uncharted Waters: Adapt and Innovate	59
Never allow yourself to merely go with the flow – Innovative ideas keep you afloat when the waters get too rough.	

Be my Tent Mate: Keep Dissidents Close   69
> *Help squash malcontents by keeping them under your radar; left alone, their negativity will grow.*

Breaking the Ice: Communicate   75
> *Don't hide behind office doors or memos – Honest and direct communication keeps everyone on the same page.*

Epilogue	83
Endurance Key Dates	85
Acknowledgements	89
Bibliography	91
Notes	97

FOREWORD

I first discovered Sir Ernest Shackleton in a *Wall Street Journal* article in 1998. I was instantly captivated by his heroism, bravery and strength – characteristics of any great survival story. I soon immersed myself in all things Shackleton -- books, movies and, in December 2006, I even retraced his steps by personally visiting Antarctica.

Some people have asked why an executive who has spent his entire career in the financial services industry was so taken with an early 20th-century Antarctic explorer. Quite simply, Shackleton's story spoke to me. At the heart of his expedition is the struggle to overcome enormous odds in the face of great adversity. I have, like many people, experienced both successes and challenges in life.

I've learned that surviving great adversity can summon some of the finest qualities we possess – or some of the worst. In Shackleton's case, becoming shipwrecked at the bottom of the earth summoned the greatest leadership skills ever found in history.

The more I read about Shackleton, the more I realized how truly heroic leadership is almost impossible to find in today's businesses. Despite all the research and programs devoted to motivating employees, most workers admit they feel disenfranchised in their daily work life. In reading the Shackleton story, it became clear to me that Shackleton's leadership lessons could benefit these very same people.

This book is my attempt to bring an extraordinary explorer's leadership lessons to those business leaders who, on a daily basis, must guide their workforce towards a common goal. Because Shackleton's story is more than just one man fighting for survival in the Arctic region – it is about coordinating teamwork under the most strenuous conditions. Even in the fast-paced and often unpredictable business world, leaders can use Shackleton's strategies to make every team effort a successful one.

In this book are inspirational lessons from one of the greatest leaders of the 20th century – lessons that can enrich both the way we work and the lives of those we lead.

Arthur Ainsberg
October 2009

INTRODUCTION

At 8:45 a.m. on December 5, 1914, English explorer Sir Ernest Shackleton and his crew of 27 men set sail from a small island at the southernmost tip of South America aboard a 300-ton wooden ship called the *Endurance*. Shackleton was leading his men to the Antarctic continent, a region known for having the most harsh and unforgiving climate on earth. Their mission: to be the first men ever to cross Antarctica on foot.

They were a determined and courageous group of explorers. And yet they would not accomplish their goal; in fact, they would never even reach the mainland. Instead, they would encounter unimaginable difficulties at every step of the way.

Paradoxically, it was this failure that would ensure their lasting fame. Instead of being a success story in the annals of polar exploration, Shackleton's expedition would go down in history as one of the most legendary survival stories of all time. And Ernest Shackleton, in his remarkable ability to keep his men alive in the face of tremendous adversity, would be hailed as one of the finest and most effective leaders of the twentieth century.

* * *

The life story of Sir Ernest Shackleton bears little resemblance to that of most modern-day managers. Born in County Kildare, Ireland, in 1874, the second of 10 children, he was an indifferent student at best, running away to sea at the age of 15. He rose in the ranks of the Merchant Marines, but his true love was polar exploration – and indeed, he had already become a world famous polar explorer by the time he undertook the *Endurance* expedition in 1914. He had dazzled the public with stories of Antarctic exploration when he returned from Robert F. Scott's *Discovery* expedition in 1904, having been one of three crewmen to come within 460 miles of the South Pole – the closest anyone had ever gotten. He beat his own record with the 1908-09 *Nimrod* expedition, this time coming within 97 miles of the Pole. The feat earned him a knighthood from the King of England, and Shackleton became an instant celebrity. He wrote a book about his voyage, gave lectures on his experiences and even received an invitation to the White House. To say Shackleton was on top of the world would be an understatement.

But it was the bottom of the world, the vast and untamed continent of Antarctica, that called irresistibly to Shackleton. He was driven not by a desire for celebrity status but by a passionate resolve to conquer the unconquerable, to be the first to gain new ground in the last uncharted region on earth. So when Norway's Roald Amundsen became the first explorer to reach the South Pole in 1911, Shackleton looked for a new quest. He set himself a goal no one had yet achieved: to cross the entire Antarctic continent on foot, from the Weddell Sea to the Ross Sea. The *Endurance* expedition was born.

Had he and his crew accomplished their goal, this book would never have been written. Because it was in their encounter with failure that a different order of success was achieved, leaving a legacy that propelled Shackleton from a footnote in history to a remarkable example of leadership.

It is this legacy that makes Shackleton's story so applicable for so many of today's managers, leaders, and aspiring leaders. Whether you are overseeing a crew of twenty-seven seamen or dozens of professional employees – whether your environment is a 300-ton wooden ship or an urban corporate office – and whether adversity comes in the form of ice floes or lower market shares, there are certain leadership qualities that are critical to the survival and success of any team, anywhere. Shackleton understood these qualities. All leaders should.

After all, leadership skills aren't only critical in life or death situations. Any project that involves a team effort towards accomplishing a goal needs a great leader to succeed. And it's only by studying the example of a leader like Shackleton that we can understand what constitutes truly great leadership. The seven lessons in leadership that comprise the second part of this book, all of them inspired by the story of the *Endurance* expedition, are lessons every one of us can use in our day-to-day efforts to support, develop, and inspire our team toward a goal.

But first, the story.

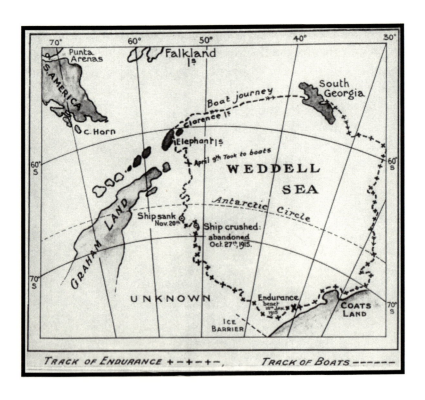

THE VOYAGE

Shackleton's first leadership challenge came long before he ever set sail: once he had decided upon his mission, he needed to find ways to finance it. In 1913, he sought the approval of government scientific societies; their support would validate the expedition as an important scientific undertaking. With World War I about to explode, however, money was tight for private enterprises like the *Endurance*. The same government that had given Robert F. Scott almost two million dollars (in today's currency) for his voyage in 1910 now granted Shackleton only half that, on condition he raise the rest himself.

Shackleton had never distinguished himself as a businessman; he tended to be attracted to business ventures that were long on big ideas and short on viability. But he understood the arts of persuasion. With help from the editor of London's *Daily Chronicle*, he made a list of the richest people in London and sent them each a detailed prospectus about his expedition. He was relying on the upper classes' penchant for vicariously enjoying the adventures of exploration, and on their particular fondness for having glaciers, promontories and lifeboats named after them.

He was not disappointed. Three donors in particular came forward so generously that their names were immortalized on his expedition's three lifeboats: Miss Janet Stancomb Willis, heiress to a tobacco fortune; Dudley Docker, an industrialist; and Sir James Jay Caird, a Scottish entrepreneur who donated a whopping $120,000 – the equivalent of $2.4 million today.

"This magnificent gift," Shackleton told a London newspaper, "relieves me of all anxiety." In addition, Shackleton received hundreds of smaller donations from individuals around the world.

Shackleton also convinced his friends who worked in the press to hype his expedition. They hailed the voyage as an opportunity for Great Britain to regain its reputation in polar exploration, and described Shackleton as the epitome of British bravery. He was able to sell the rights to any profitable aspects of the expedition, including its photography, movie rights and a book.

He even arranged to give speeches about his experiences when he returned. It is proof of his persuasive powers that those who bought the rights to the expedition's commercial properties simply took for granted that Shackleton would come back alive!

Shackleton's determination and charisma paid off. He raised over $10 million in today's currency, and his dream of an *Endurance* expedition became a reality.

As no one had ever imagined a way to walk across the Antarctic and survive, it was up to Shackleton to invent one. His strategy was simple: he would employ two ships, one to approach the Antarctic by the Weddell Sea, to the east, and another to reach it from the west, by the Ross Sea. The *Endurance* would sail through the Weddell Sea and drop off six men, who would travel across the Antarctic continent by dog and motor sleds until they reached the Ross Sea. Shackleton calculated 100 dogs would be needed for the trek across Antarctica. The *Aurora* would leave men on the Ross seacoast, who would walk inland and drop off supplies along the second half of the *Endurance* group's route. The *Aurora* would remain at the Ross Sea coast until the traveling crew arrived. Some called the plan audacious – and it was. But Shackleton was confident that it was sound.

"[T]he first crossing of the last continent should be achieved by a British Expedition," Shackleton declared about his plans. "[W]e were taking part in a strenuous campaign for the credit of our country."

The *Endurance* sailed out of Plymouth on August 8, 1914, just as war clouds gathered over Britain. The ship stopped at Buenos Aires before reaching its last port of call, the sub-Antarctic South Georgia Island, on November 5. Before leaving Buenos Aires, Shackleton had fired several drunk and disobedient crewmen.

It was, as Orde-Lees wrote in his diary, "splendid having Sir Ernest on board. Everything works like clockwork & one knows just where one is."

On the December day when the *Endurance* set sail from South Georgia Island and headed for Antarctica, whalers on the island warned Shackleton that ice conditions in the seas were worse than usual. They were right. On January 18, 1915, a little over a month after leaving South Georgia, the *Endurance* was frozen in the ice of the Weddell Sea. The expedition was a day's sail from reaching their intended landing spot, Vahsel Bay.

"The land showed faintly to the east," Shackleton recalled. "The ice was packed heavily and firmly all around the *Endurance* in every direction as far as the eye could see from the masthead. There was nothing to be done till the conditions changed, and we waited through that day and the succeeding days with increasing anxiety."

The crew tried everything they could to free the ship from the harsh grip of the polar ice, chipping away with chisels and picks to make room for the ship to ram at full speed into the floes ahead. It was no use. On February 24, Shackleton declared the *Endurance* a winter station. They would have to live on board the ship and wait for the Antarctic spring, when higher temperatures might thaw the ice. Unfortunately, spring was nine months away.

As the permanent darkness of the polar winter set in, Shackleton and his men could do nothing but shore up their supplies of seal meat and watch helplessly as the ice floes they stood on slowly floated further away from the Antarctic mainland, widening the distance between the crew and their destination. To make matters even worse, they had absolutely no way to contact the outside world for help. It would be another 10 years before radio technology would make it possible to communicate from the desolate Arctic region.

"Shackleton at this time showed one of his sparks of real greatness," Macklin wrote in his diary. It was a scenario ripe for depression, unrest and friction among the crew members, but Shackleton made it his mission to keep his crew unified and positive. He encouraged games, skits, and music to help them through the long months of the Antarctic winter night. Remarkably, the men drew closer together and morale remained generally high. This wasn't Shackleton's main goal when he set out towards Antarctica in 1914, but it would become his most pivotal.

In early August 1915, storms began to pick up outside the *Endurance*. Ice floes crashed into each other, once resulting in a 20-

foot high fold of ice. Other floes took hold of the ship, cracking its beams. By October 27, the frozen assault proved too brutal for the ship to withstand. Amid the cracking ice, moaning wooden planks and howling dogs, the men were ordered to abandon ship. They set up a temporary campsite nearby, christened the "Dump Camp," leaving most of their supplies stored in the deserted *Endurance*.

But Shackleton didn't wallow in despair. "The disaster," Shackleton later recalled, "had been looming ahead for many months, and I had studied my plans for all contingencies a hundred times."

He proposed to the crew his plan to walk northwest across the ice floes and back to land, dragging sledges filled with vital supplies and the lifeboats as well, in case they came to open water. He also used the opportunity to show appreciation for his crew's hard work.

"I thanked the men for their steadiness and good *morale* they have shown in these trying circumstances," Shackleton later wrote, "and I told them I had no doubt that, provided they continued to work their utmost and to trust me, we will all reach safety in the end."

Such direct and open discourse was crucial for the group of downtrodden men. As one crewman wrote, "the job was now up to us…We were in a mess, and the Boss was the man who could get us out."

Insisting that it was imperative to travel light, he set an example by placing his gold cigarette case and Bible in the snow. Queen Alexandria had given Shackleton the Bible at the start of the voyage. For him, the most valuable part of the book was the inscription she had written on the first page: "May the Lord help you to do your duty & guide you through all the dangers by land and sea. May you see the Works of the Lord & all His Wonders in the deep." As the crew shed extra weight in preparation for their arduous march, Shackleton made sure to rip out the page with the Queen's words to carry with him. If he were to stay focused on his goal of getting everyone home alive, inspirational words like these were invaluable.

"A man under such conditions needs something to occupy his thoughts, some tangible memento of his home and people beyond the seas," Shackleton would later write when listing the items items he saved.

One of these items included the banjo: "I took it with us as a mental tonic," Shackleton recalled. "It was carried all the way through with us…and did much to keep the men cheerful." Shackleton understood he wasn't the only one who needed tools for encouragement; if he expected his men to follow him, they would need to be inspired too.

It was now spring in Antarctica, with temperatures rising into the 20s (Farenheit) – and making travel even more treacherous than before. Dragging the one-ton boats through wet, heavy slush was a torturous process; continuous snowstorms further hampered progress, and the unstable ice packs were cracking in every direction. It did not take long for Shackleton to realize that they were covering less than a mile a day. He halted the journey, deciding that their best hope would be to wait until the ice floes broke apart and they could launch the lifeboats toward land. They set up "Ocean Camp" and began regular salvage trips back to the ship to retrieve as much of their stores as possible.

On the night of November 21, 1915, the men watched from a distance as the ice beneath the *Endurance* broke apart. The ship rose under the heaving ice floe – and then sank rapidly into the black, frigid waters and disappeared. Their home and their source of stored supplies was gone.

"It gave one a sickening sensation to see it," one crewman wrote in his diary, "for, mastless and useless as she was, she seemed to be a link with the outer world."

Shackleton's crew had shown astonishing reserves of spirit for more than a year. Now, with the ship gone and very little to do, they began to grow restless. So on the day before Christmas, 1915, Shackleton initiated a second try at a journey across the ice. He understood the importance of getting the men moving again.

"It will be much better," Shackleton later wrote, "for the men in general to feel that, even though progress is slow, they are on their way to land than it will be simply to sit down and wait for the tardy northwesterly drift to take us out of this cruel waste of ice."

Once again they set out, dragging supplies and lifeboats, across the sodden, cracking ice. The going was so slow that at times it took two hours to go less than a thousand yards, but Shackleton's plan was working.

"All hands were very cheerful," Shackleton noted, "a relief from the monotony of life on the floe raised all our spirits."

But at the rate they were going, it would have taken the crew about three hundred days to reach land, and their lifeboats, which were taking a beating, might not even have survived the journey. And so once again Shackleton called a halt and set up camp. This time they had no choice but to stay put until the ice floes broke and they could sail their lifeboats on open water.

The new camp was aptly named "Patience Camp." Without the supplies and small comforts that had buoyed the men's spirits before, there was little to do but wait and hope the northward drift would slowly bring their ice floe closer to land. The weather was relentlessly bleak, with continual blankets of rain and fog. Heavy gusts of ice and wind sometimes made it difficult even to stand up. Food was largely limited to seal meat, which Shackleton noted, prevented "a single case of scurvy…amongst the whole party, yet it was a badly adjusted diet, and we felt rather weak and enervated in consequence."

Perhaps worst of all was the acute boredom. As one of the crew wrote in his diary, there wasn't much to do but "get into our sleeping bags. And smoke away the hunger."

For the first time Shackleton had the terrible sense that he had no real control over the fate of his men. On January 26, he wrote in his diary: "No news…Waiting. Waiting. Waiting." He kept his feelings to himself and his eye on the ice, waiting for the perfect moment when the floes would break apart completely and they could sail to safety.

All hearts lifted when at last the ice floes began to crack. By April 9, 1916, there was enough open water for the crew to launch the three lifeboats and head west toward land. Deception Island, just west of Elephant Island, would provide them with the best chance of discovery by passing whalers.

The polar seas were anything but smooth sailing. Waves bearing huge chunks of ice crashed up against each other, pummeling the boats and nearly crushing them. Ice showers covered the boats and men alike with a frozen crust. Killer whales surrounded them in all directions. Worst of all, the winds that assaulted them were forcing the boats eastward, away from land.

Everything that could be spared was thrown overboard, and the boats were turned around and headed once more into the gale. This time they set their sights for Elephant Island, the closest piece of land.

On April 15, Shackleton and his men staggered onto the shores of Elephant Island. They had been traveling in lifeboats for six days. Exhausted, injured, and frostbitten, they collapsed into sleep.

It had been a year and a half since their feet had touched land.

They soon discovered that Elephant Island was no paradise. The island was strewn with penguin dung and constantly swept by blizzards and gale winds. The men tried to get some rest after pitching their tents, but sleep did not come easy. A blizzard blew in and ripped apart one of the larger tents and deflated some of the others. Sleeping bags became soaked with precipitation. Supplies were blown away by the wind. But even greater problems were on the horizon. While the men enjoyed blubber and seal for food, Shackleton knew they only had five weeks' worth of rations left. The polar winter was looming, and increasing ice would soon prevent them from hunting the seals and penguins that passed the island.

Perhaps even more troubling were signs of demoralization growing amongst the crew. "They were disinclined to leave the tents when the hour came for turning out," Shackleton wrote, "and it was apparent they were thinking more of the discomforts of the moment than of the good fortune that had brought us to sound ground and comparative safety." Their future looked bleak, and Shackleton knew the situation called for immediate action.

"A boat journey in search of relief was necessary," Shackleton wrote, "and must not be delayed. That conclusion was forced upon me."

He decided to strike out for South Georgia Island, some 800

miles away, to seek assistance. It was an ambitious goal, a feat which normally would have been considered far too dangerous if not for the risks the men faced if they remained on Elephant Island. He chose five men to come with him: James Worsley, Thomas Crean, Harry McNeish, John Vincent and Timothy McCarthy. The *James Caird*, the third lifeboat, was repaired and readied for the journey. To preside over the remaining crewmembers on Elephant Island during his absence, he appointed Frank Wild as leader.

Initially, Wild had asked Shackleton if he could join him on the boat journey to South Georgia, but Shackleton refused. "I told Wild at once that he would have to stay behind," Shackleton later wrote. "I relied upon him to hold the party together while I was away…I trusted the party to him." By appointing his most trusted crewmember as a surrogate leader while he was away, Shackleton was able to focus his energies on the next critical task: reaching South Georgia.

On April 24, 1916, Shackleton and Wild shook hands in farewell, and to a chorus of weak hurrahs from shore, the *James Caird* – a 22-foot-long and 6-foot-wide whaler – set sail. "There was hope in their hearts," Shackleton recalled about the men he left behind, "and they trusted us to bring the help they needed."

The first day at sea provided good sailing weather, but then conditions worsened rapidly. Icy blasts of wind hurled the *James Caird* this way and that across the waves. Ice continually froze over the boat and had to be chipped away again and again. The men pumping the boat had numb hands from holding the brass pump underwater; the men at the wheel were subject to a constant shower of freezing water. At one point a crewman became frozen in place, and the others had to help him just to get into his sleeping bag.

Exhausted though they were, sleep didn't come easily. Their cramped sleeping quarters in the foremost point of the boat's hull could only be reached by crawling over the rocks that served as ballast at the bottom of the boat. Even when they had attained their sodden sleeping bags, the violent pitching of the lifeboat made sleep nearly impossible. And once they fell asleep, they might be wakened to bail frantically when the boat was engulfed by a giant wave.

After fourteen days at sea they opened their second keg of drinking water – only to discover that it was brackish and barely drinkable. Worst of all was the fear that they had missed the island altogether and were headed into the 3,000 miles of open sea that lay between them and Africa.

On May 8, fourteen days after setting off from Elephant Island, land was sighted. But almost immediately a tremendous storm arose, threatening to dash them against the rocky shore they were trying to reach. When at last they attained land safely two days later on May 10, the men dropped to their knees by a nearby stream to drink much-needed water. "It was a splendid moment," Shackleton would later write, "…the pure, ice-cold water…put new life into us." They didn't realize it then, but the 16-day, 800-mile journey they had just completed would go down in history as one of the most extraordinary open boat voyages in sea-faring history.

Their relief was quickly tempered by the realization that they were on the other side of the island from the whaling station. Attempting another journey on the battered and crippled *James Caird* was out of the question. Exhausted, starving and frostbitten, they would have to cross the 6,000-foot glacial mountains of the island on foot. No man had ever crossed South Georgia before, and no maps existed to guide them in their journey. They would have to rely solely on the same tenacity and courage that got them to South Georgia in the first place.

"Over on Elephant Island," Shackleton mused, "twenty-two men were waiting for the relief that we alone could secure for them. Their plight was worse than ours. We must push on somehow."

Shackleton decided that McNeish, Vincent and McCarthy were simply too ill to attempt the trek. They would stay behind on the coast, while Shackleton, accompanied by Worsley and Crean, set out towards Stromness whaling station. After resting in a cave for a few days, the three men embarked in the early hours of May 19.

The interior of South Georgia Island was a terrain of glacial mountains, steep gorges and steeper cliffs. Again and again, the men struggled to reach the top of a mountain only to find that the other side was a sheer cliff face, and they had to retrace their steps.

After 23 hours of walking, the men thought they had reached the valley of Stromness Bay, but quickly realized this was not the case. Having taken a wrong turn earlier, they again had to turn and climb back the way they had come.

The next morning, they came across the hills that surrounded Stromness whaling station. In the distance, they could hear the 7 a.m. steam whistle. After sliding down a 500-foot icy slope, the men soon found themselves pushing through frigid waters that abruptly ended in a waterfall. They fastened themselves with rope they attached to a large boulder, and slowly lowered themselves down the 25-foot drop.

At 4 p.m. on May 20, 1916 – nineteen months after first setting off from South Georgia Island – Shackleton, Worsley and Crean, exhausted both physically and emotionally and given up for dead by the entire world, arrived at Stromness whaling station. It had taken them nearly 37 hours of constant trekking to cross South Georgia, but after nearly a year and a half in peril, they had finally reached civilization.

The whalers were quick to help the three battered and fatigued figures that stumbled into Stromness station. They had heard of Shackleton, but did not recognize him as the "funny-looking" man that now stood before them. One older whaler, Shackleton wrote, "started as if he had seen the devil himself." Another whaler later recalled his emotion when Shackleton identified himself by name. "Me – I turn away and weep," he said. "I think manager weep, too."

The three men were given hot baths and plenty of food, and then Worsley was dispatched in a relief boat to gather McNeish, Vincent and McCarthy on the other side of the island. Shackleton had only one thought in his mind: to save the men left behind on Elephant Island. He was aware that in a strange way these men had helped him to save himself. "It might have been different," he later recalled, "if we'd had only ourselves to think about. You can get so tired in the snow, particularly if you're hungry, that sleep seems just the best thing life has to give...But if you're a leader, a fellow that other fellows look to, you've got to keep going."

Three days later Shackleton was on the whaleboat *Southern Sky* along with Worsley and Crean, heading back to Elephant Island. One hundred miles from land, the ship ran into ice; they pushed on for forty more miles but nearly got stuck again. They were forced to return to land – this time to the Falkland Islands to find another boat that could handle the pack ice.

Shackleton contacted the King and Admiralty in London for assistance, asking for one of Robert F. Scott's old ships. But the war in Europe meant there were not many ships to spare. With the British government slow to respond, Shackleton approached other governments, and Uruguay offered its *Instituto de Pesca No. 1*. By June 10 he was once again en route to Elephant Island. But the ice still wouldn't budge. Close enough to Elephant Island that he could see it, Shackleton had no choice but to turn back.

The government of Chile informed him that they could have a sturdier ship ready for him by September 10. Shackleton became increasingly irate and desperate. He was going gray, Worsley noted; he was beginning to drink. He knew that time was running out for his marooned crew – if it had not run out already.

The Chilean government came through on August 25th with a naval ship, the *Yelcho*. This time Shackleton was able to push through the ice. On August 30, 1916, he arrived at Elephant Island. It had been four months since he had left.

Every one of the twenty-two men was still alive.

After saving his *Endurance* crewmen, Shackleton immediately turned his attention to the Ross Sea Party, who had traveled on the *Aurora*. Though the ship had reached Cape Evans, Antarctica, and several men from the crew began laying supplies in anticipation of Shackleton's trek across the continent, the *Aurora* blew out to sea on May 7, 1915. Ten crewmen were left stranded on Antarctica without supplies, most of which still remained on the ship. Attempts to contact the men on shore failed. The boat, with other crewmen still on board, drifted in pack ice for 10 months before making land at New Zealand on April 2, 1916.

Shackleton, intent on taking command of the *Aurora* and rescuing the remaining Ross Sea Party crewmen at Cape Evans, sailed to New Zealand in December 1916. But before he had made contact with the outside world, the governments of Australia and Britain had made arrangements for a relief effort to repair the *Aurora*. Shackleton's old friend, Captain John K. Davis, had been placed in charge of the ship. Shackleton, satisfied with the relief effort, joined Davis on the trip to rescue the 10 remaining crewmen. Upon arrival at Cape Evans, Shackleton learned only seven of the 10 crewmen had survived. These three men represented the only fatalities of the whole expedition.

Shackleton took full accountability for the lost lives. "I [am] Commander of the Expedition," he would later write, "[and] though I was thousands of miles away, the responsibility still lay on my shoulders."

In the spring of 1917, Shackleton and his men came home to an England and world gripped by war. New and more lethal forms of fighting, including poison gas and submarine warfare, were claiming the lives of millions of people. It was, Shackleton realized, "the most stupendous war in history."

With the war came a new definition of hero. The public was now less likely to celebrate survival stories like Shackleton's than to mourn soldiers who had heroically given their lives to the war effort. There was little fanfare for the returning *Endurance* expedition.

Shackleton had always felt guilty for leaving England at the brink of war, and immediately volunteered for the war effort. He traveled to South America to increase support for Britain and then to Russia, where he helped transport troops and equipment. Many of the men from the crew of the *Endurance* went on to serve on the front lines of the war.

After the war ended, Shackleton turned his attention back to polar exploration. He embarked on his final expedition in 1921, aboard the *Quest*. His aspirations were less lofty than before: the *Quest's* mission was to chart the Antarctic coastline and explore islands in the area. Eight men from the *Endurance* joined the 18-man crew.

The *Quest* arrived at South Georgia Island on January 4, 1922, but Shackleton never made it ashore. On January 5, before even leaving the ship, Shackleton died of a heart attack. He was 47 years old. His crew buried him on South Georgia Island.

Sir Ernest Shackleton never attained the South Pole. He never succeeded in walking across the Antarctic continent or in charting its coastline. If we were to judge him by his success in meeting those goals, he would come up short.

Shackleton's legacy lies in his extraordinary achievement of a different kind of success: His ability to inspire his men with a vision; to enable them to work effectively as a team under conditions of unthinkable difficulty; and, finally, to keep them alive. His genius, quite simply, was in leadership.

THE LESSONS

Sir Ernest Shackleton

LESSON ONE

The Fire Within:
Feel the Purpose in Your Gut

> "[Shackleton] had qualities which produced a ready response...His great passion was an inordinate personal ambition which knew no limits and sometimes...soared...beyond the physical efforts of which he was capable." – *Eric Marshall, Nimrod expedition*

Everyone knows the importance of having a clear and ambitious goal. Shackleton's original mission was certainly clear and ambitious – he was specific in his undertaking and motivated to accomplish something no one else had done.

But that's not the same as feeling a burning purpose in your gut. This kind of feeling goes deep within your heart and soul to the very core of your existence. It keeps you up at night and stays with you every moment of the day.

This is how Shackleton felt about his new purpose after the *Endurance* sank beneath the frigid waters of the Weddell Sea. His original goal to cross Antarctica was replaced with a different and much more urgent mission: to get everyone home alive. It was more than an obligation; it was the only way he could live with himself. He had brought these men into this mess, and he was going to get them out of it.

"The task now was to secure the safety of the party," Shackleton would later write, "and to that I must bend my energies and mental power and apply every bit of knowledge that experience of the

Antarctic had given me. The task was likely to be long and strenuous, and an ordered mind and a clear program were essential if we were to come through without loss of life."

Throughout the rest of the journey, he paid close attention to the activities and health of his crewmen, knowing that one wrong move could cost a life. As one crewman wrote, Shackleton "invariably raises hell if anyone gets injured in any way unnecessarily." When Thomas Orde-Lees ventured too far on the ice on his bicycle, Shackleton forbade him from ever using it again. When Dr. Alexander Macklin and Lionel Greenstreet decided to "ride" a little ice floe that was floating in open water, Shackleton caught them in the act, and one hard look from him was enough to teach them to never do it again. He would not tolerate frivolous activities that could undermine his abiding purpose, which was to bring his crew home safe.

"His first thought," Greenstreet, the ship's First Officer, said, "was for the men under him. He didn't care if he went without a shirt on his back so long as the men he's leading had sufficient clothing. He was a wonderful man that way; you felt that the party mattered more than anything else."

Privately, Shackleton concurred. "I pray God," he wrote in his diary, "I can manage to get the whole party to civilization."

Shackleton's capacity to sustain a burning passion had been with him all his life. When he went to sea at fifteen, his four years of maritime apprenticeship were a miserable experience. He endured backbreaking labor, violent storms, and the coarse behavior of his fellow crewmen; he wrote hundreds of letters home. But he was never tempted to quit. He never lost sight of his passionate dream of becoming the captain of a ship. It was more than a plan – it was a constant, burning purpose.

"I wanted to be free," Shackleton later recalled. "I wanted to escape from a routine which didn't at all agree with my nature and which, therefore, was doing no good to my character. Some boys take to school like ducks to water; for some boys, whether they take to it or not, the discipline is good; but for a few rough spirits the system is chafing, not good, and the sooner they are pitched into the world, the better. I was one of those."

His plan paid off. He obtained his master's certificate in 1898 and by 1900 had his sights set on Antarctic exploration.

Years later, when he faced extreme hardship and catastrophic misfortune with the *Endurance* expedition, his capacity for sustaining a burning sense of purpose stood him in good stead. He also knew the importance of keeping sources of inspiration close to him at all times, to help shore up his resolve. Literature and poetry books were constant companions on his expeditions, and he turned to passages from Shakespeare and poet Robert Browning to help himself stay focused and motivated. After all, he would later write, "[l]oneliness is the penalty of leadership."

Shackleton's devotion to his crew was a given. But when confronted by one obstacle after another, he needed to make sure his spirit did not falter. Shakespeare, Browning, the words of his Queen in the Bible – these things nourished Shackleton's spirit and spoke to his soul. They helped him hold fast to his sense of purpose, even in the most desperate of circumstances.

* * *

In our modern corporate world, goals are far too often built strictly around increasing profitability and the bottom line. While these are important ambitions for keeping a company successful and efficient, they don't often inspire a burning sense of motivation. When a leader is guided by a driving personal vision, the company as a whole is energized, invigorated and focused. Without that vision, there is a loss of energy and team spirit throughout the ranks.

Take, for example, Gary Erickson of Clif Bar & Co. Founded in 1992, Clif Bar has risen to become one of the leading brands in natural energy and snack bars, achieving $200 million in sales in 2008. The company was initially driven not to earn a profit but to create tasty energy bars that could sustain Erickson and his friends during long bike races and rock-climbing trips. As an avid athlete, Erickson was unhappy with bars already on the market and decided he "wanted to make a better energy bar for friends and myself." This "better" bar would be delicious, nutritional and provide the energy needed during strenuous physical activity. Erickson spent six months in his mother's kitchen perfecting the bar before premiering it at a bike show in September 1991. It created an instant buzz.

Erickson's passion and vision transformed a relatively simple idea into a profitable success. Early Clif Bar employees shared Erickson's vision of an environmentally-friendly and community driven business and stuck with the company as it grew. His passion also kept him focused when he was presented with a $120 million buyout of Clif Bar in 2000. Believing the naysayers that said the company would eventually lose steam and fold under larger competitors, Erickson was literally minutes from signing away Clif Bar when he realized when "you sell your company you sell your vision." He realized that with the hectic schedule leading up to the buyout, he hadn't been listening to his gut, which told him he wasn't through with Clif Bar just yet. Fearful that a buyout would compromise the integrity and future path of Clif Bar in its remaining years, Erickson refused to sell.

"Operating from the gut or intuition isn't about making random or illogical choices," Erickson would later say. "It's about being able to bring experience, logic, passion and creativity to bear on the unknown, and, in a split second, make sense of it."

Employees would later say that the passion began to die at Clif Bar during the time leading up to the company's sale. The potential selling of the company sent the message that it was a "big dog eat little dog world", and that Clif Bar didn't believe in itself enough to compete. By taking control of the reigns and refusing to sell, Erickson single handedly revitalized the company. Erickson's concern that Clif Bar couldn't compete with larger companies was transformed. He began encouraging employees to believe in the company and reemphasized the idea that delivering high-quality products was more important than cutting production costs – their products must be something they could all believe in. It worked. In under three years, Clif Bar grew from $40 million to $100 million in annual sales. Its Luna Bars overtook PowerBar as the number one bar in the country in grocery and outdoor channels. Erickson was even revitalized personally – he slept better, began to ride his bike again and his home life improved.

Over 15 years since it inception, Erickson continues to lead Clif Bar with new products and a sustained dedication to the environment and company vision – even being named "Best Boss" by Fortune Small Business in 2003.

Like Erickson, environmental activist Gary Hirshberg represents another figure who transformed his passion for healthy foods and the environment into a profitable business. Hirshberg was inspired to start Stonyfield Farm, the world's leading organic yogurt-maker, after watching a Kraft Food presentation at Walt Disney's Epcot Center in Orlando, Florida, in 1982. Disturbed that Kraft touted "future farming" as growing food in plastic tubes and in buildings heated and cooled by fossil fuels and chemical fertilizers, Hirshberg had an epiphany: "I have to become Kraft." Hirshberg realized that only a company with equal financial clout as Kraft could redefine the terms of sustainable practices to save the planet – and get people to listen.

At the time, Hirshberg was an executive director at an ecological research and education center in Cape Cod, Mass. Though proud of the environmental gains being made at the center, it was no business academy. Yogurt was sold only to fund the school, and the leaky barn and several cows couldn't do much more than that. To "become Kraft," as Hirshberg vowed, he would have to create a much larger undertaking. Accordingly, he and Samuel Kaymen soon bought farmland 30 miles from Cape Cod and launched Stonyfield Farm in 1983.

One of the first things Hirshberg did for his new venture was to write a mission statement. His five-point plan included providing the highest-quality and all-natural, certified organic dairy products; educating consumers and producers about the value of protecting the environment; and serving as a model that economically and socially responsible businesses can also be profitable. Though Hirshberg admittedly had doubts whether Stonyfield Farm could stay afloat by rejecting conventional business wisdom, his mission statement helped prioritize the goals for Stonyfield Farm, and reminded the company why it was created in the first place.

Hirshberg's ideas worked. Stonyfield Farm has since expanded into the third largest yogurt company in the United States and the world's leader of organic yogurt, generating $300 million in revenues in 2007. Its products also include smoothies, ice cream and milk. In turn, Hirshberg returns 10 percent of the annual profits to environmental groups. He also pays more for organic ingredients, supports family farmers and helps conventional farmers become organic, an expensive process.

Through these practices, and paying the price to cut back its energy consumption, Stonyfield Farm's products cost more than its competitors, but consumers keep buying. The company's environmental values have created a loyal consumer base. Hirshberg's passion to prove that big businesses *can* operate in green, eco-friendly ways without sacrificing its success has helped lead it into the flourishing company seen today.

Also consider the example of Steve Jobs, whose vision – as passionate as it was brilliant – helped to drive the founding of Apple, Inc. When Jobs left the company in 1987, Apple began to lose the clarity of purpose and creative energy that had distinguished it from the beginning. He returned a decade later to a company needing his enormous creative energy and skill-set. At the time, Apple was only six months from bankruptcy.

With Jobs back at the helm once more, Apple's recovery was swift. Jobs cut unnecessary fractions of the company. He made it a mission to remind people what Apple stood for: imaginative, elegant, and high-quality products. He surrounded himself with creative thinkers and helped redefine Apple through the "Think Different" campaign, which celebrated individuals who pushed the human race forward. In 1998, as result of his efforts, what would be the first of many innovative products was unveiled to the world: a bondi-blue iMac. It would mark the beginning of a huge resurgence for Apple. Jobs's personal vision and his burning, take-no-prisoners quest to realize it – quite simply, made all the difference.

To discover and articulate a meaningful purpose entails a bit of soul-searching. It is not likely to emanate from a Board of Directors, and it probably shouldn't contain words like "banner year," "profit margin," or "global initiative." If it sounds like jargon, chances are it's not driven by your own authentic personal vision. It should come instead from a place deep within you, a place that pre-dates ascending the corporate ladder and will continue to exist long after retirement and your 401K. It should come from the heart. You should feel it in your gut. And, sometimes, you may have to take a risk in order to pursue it.

Howard Schultz, now CEO and Chairman of Starbucks, gave up a prestigious job in New York to move 3,000 miles away and become part of a small chain of coffee stores in Seattle. His mother begged him not to. But the first time Schultz walked into a Starbucks Coffee, Tea and Spice store in Seattle's historic Pike Place Market district, he was mesmerized by the full flavor of its coffee and the unique culture of roasting coffee beans. He immediately envisioned a growth for the store that would spread it into a nationwide chain, thus sharing its passion and ability to make quality coffee with the entire country. "It wasn't until I discovered Starbucks that I realized what it means when your work truly captures your heart and your imagination," Schultz writes in his book, *Pour Your Heart into it*.

At first, Starbuck's owners, Gerald Baldwin and Gordon Bowker, turned down Schultz's offer to transform the small chain into a nationwide success. It was not what they envisioned for their humble company, and they were worried Schultz would clash with the coffeemaker's existing culture. But Schultz didn't accept no for an answer. After an intense conversation with Baldwin in which Schultz laid out his conviction and passion to turn Starbucks into something big without compromising its integrity, Baldwin hired him.

Starbucks's widespread success wasn't immediate – In fact, Schultz left Starbucks to start his own chain of coffee shops in 1985, only to buy out Starbucks when its owners offered it for sale in 1987. But Schultz re-invented Starbucks into the nationwide success it is today. It has become synonymous with great-tasting coffee and the best ingredients. It introduced many Americans to exotic drinks that had never been commercially available in the United States, like espressos. Schultz intentionally wanted to mimic the espresso bars he loved in Italy, where a daily stop for a café latte was part of every

Italian's day. It also provided an inviting, social atmosphere, where patrons could take a break from their busy day and relax. Similar to French cafes and German beer gardens, Starbucks, in its own way, provided a sense of community outside work and home, and became a place where people could meet.

What would have happened if Schultz had accepted Starbuck's initial decision not to hire him? He'd likely still have a snug job, but he would lack the drive and passion Starbucks ignited in him. A good purpose is one worth fighting for. Sometimes you don't have to take no for an answer.

So how will you know if you've found it? You and your team will be galvanized, focused and in sync with one another. Your passion about the mission will inspire them to be passionate about it too. And if you are fortunate enough to achieve your goal, *don't* become complacent – set new goals, reach even higher! This kind of thinking could have helped Ford retain its spot as automobile leader in the early 1900s.

In 1907, Henry Ford unveiled his bold idea to "democratize the automobile." He vowed "to build a motor car for the great multitude…It will be so low in price that no man making a salary will be unable to own one…everyone will be able to afford one, and everyone will have one."

Though the Ford company was one of 30 companies fighting to be a leader in the emerging automobile industry, Ford's ambitious goal inspired the design team to work until eleven at night in order to create the kind of car Henry Ford envisioned. Charles Sorenson, a member of the design team, recalls that he and Ford once worked for 42 hours without taking a break.

The hard work paid off. In 1908, Ford introduced the Model T, also known as the "Tin Lizzy." In its first year of production, Ford sold more than 10,000 cars, and another 18,000 in its second year. While one of its competitors, General Motors, saw its market share dip from 20 to 10 percent, Ford climbed to number one in the automobile industry.

But instead of raising the bar even higher, Henry Ford became complacent. He wouldn't acknowledge that the public was soon demanding larger, faster and more stylish cars. As a result, in the 1920s, there was a sudden decline in demand for the Model T. Meanwhile, Alfred G. Sloan, Jr., who now managed General Motors, instituted at his company new business strategies and a corporate culture with an emphasis on efficiency. Perhaps his most brilliant move was to allocate a specific price range for different divisions of General Motors vehicles, from the Chevrolet to the Cadillac. By keeping prices at the high-end of each division close to the price at the low-end of the next division, Sloan could convince buyers to spend a little bit more for the added status of driving a car in a more "prestigious" division. In 1923, the company began making major cosmetic revisions to its automobiles every three years. Soon GM's Chevrolet surpassed the Model T as the most popular passenger vehicle.

On May 27, 1927, Henry Ford ended production of the Model T, and General Motors had surpassed Ford as an industry leader. Though Ford had mastered his initial purpose, he hadn't created another one in its place. Don't allow yourself or your team to slip into this trap; while you can celebrate a job well done for reaching your goals, remember to reestablish a new purpose before you become too comfortable.

And one more thing. Nourish your spirit, as Shackleton did, by finding sources of inspiration and renewal. Read. Take a Yoga class. Sign up for painting lessons. Travel. Listen to music. Keep a journal. Or, like Clif Bar's Erickson, go outside and take long walks or bike rides to clear out the mental clutter in your mind. It was during a much-needed walk that Erickson listened to his heart and decided to back out of the deal to sell Clif Bar. The more you are inspired in a variety of ways, the better you will be at holding fast to your vision.

Second-in-Command, Frank Wild

LESSON TWO

A Captain Needs a First Mate: Choose a Powerhouse for Your Number Two

"It's the old dog for the hard road every time." - Shackleton

Competition is the name of the game in the business world. We compete fiercely at every level for career advancement, whether for reasons of financial gain, personal satisfaction, or, usually, both. If you are in a leadership position, it may be hard to place your trust in someone just as good as (or maybe even better than) you are... especially if that person would love to have your job.

But the importance of having a competent Number Two by your side cannot be underestimated, even if you are a strong and capable leader. Shackleton understood that his second-in-command would need to be a strong and reliable source of support, and in fact might have to take over his role at some point, either temporarily or permanently. And he knew that this person needed to have skills and competence that mirrored and matched his own.

He found these qualities in Frank Wild, who he praised for his "energy, initiative, and resource" and "quick brain." Born in Yorkshire, England, Wild had joined the Merchant Navy at the age of sixteen before transferring to the Royal Navy. He was an experienced Antarctic explorer; he had worked alongside Shackleton on the *Discovery* and later on the *Nimrod,* and had accompanied

Douglas Mawson's 1911-1913 scientific expedition to Adelie Land in Antarctica as a sledging expert. As the only member of the *Endurance* crew to have sailed on all three of Shackleton's expeditions, Shackleton could be sure that he had the skills and competence required.

Just as important, Wild embodied the two personal qualities Shackleton valued most: loyalty and optimism. Loyalty was a paramount factor in the choice of a Number Two. All leaders need to be absolutely certain that their second-in-command will remain faithful to his or her mission, plans and directives. Anything less could undermine your authority over the team and potentially cause a rift in team unity. Shackleton was convinced of Wild's capacity for steadfast loyalty. And he was right.

Wild's "unfailing optimism" and friendly disposition were equally valuable. Shackleton sometimes used Wild as a bridge between himself and his men, so it often fell on Wild to communicate arduous or grueling directives. Crewman Orde-Lees wrote in his diary: "He acts as Sir Ernest's lieutenant and if he has any orders to give us he gives them in the nicest way, especially if it is instructions to carry out some particularly nasty work..." And ship's doctor Macklin noted, Wild was "calm, cool or collected, in open lanes or tight corners he was the same; but when he did tell a man to jump, that man jumped pretty quick. He possessed that rare knack of being one with all of us, and yet maintained his authority as second-in-command."

With such a high-caliber Number Two man, Shackleton doubled the managerial talent overseeing his men; in effect, the *Endurance* crew had not one but two gifted leaders. When they spent their first night sleeping on the ice after abandoning their ship, it was Wild who was by Shackleton's side, keeping watch over the men in the bitter cold. The next morning, it was Wild who assisted Shackleton in visiting all the tents and serving hot coffee.

Throughout the voyage, whenever Shackleton needed to concentrate on a specific situation or crewman, he knew that the rest of the group would be well supervised. And when he sailed to South Georgia for help, leaving Wild in charge of 22 crew members on

Elephant Island, he could be certain that Wild would undertake the mission of keeping them alive with a zeal that matched his own.

"It is largely due to Wild," Shackleton would later write about the survival of the crewmen on Elephant Island, "that the whole party kept cheerful all along, and, indeed, came out alive and so well…His cheery optimism never failed…I think without doubt that all the party who were stranded on Elephant Island owe their lives to him."

* * *

In the business world, having a strong second-in-command can be crucial to the success of your enterprise. Consider John Francis "Jack" Welch, the strong-willed, charismatic leader of General Electric for 20 years. Under his management, the value of GE increased from $13 billion to several hundred billion. Welch was famously interested in entrusting key positions to outstanding, assertive individuals. He prized above all what he called "E to the fourth power" – that is, the ability to demonstrate enormous personal energy, to motivate and energize others, to be energetically competitive, and to execute plans with energy. In surrounding himself with people who demonstrated "E to the fourth power" – and in particular, appointing the rising young star Jeffrey Immelt as CEO of GE Medical Systems – Welch ensured that his company would be infused with the brain power and initiative it needed to flourish. Other notable stars included Gary C. Wendt and Robert C. Wright. Wendt led GE Capital Corp. to exceptional heights when it contributed to nearly 40 percent of the company's total earnings, and Wright managed an astonishing turnaround at NBC, leading it to a fifth straight year of double-digit earnings gains in 1997 and a No. 1 position in prime-time ratings.

Welch placed so much importance on his executive talent, he began thinking about his successor years before his retirement. "From now on," Welch said in 1991, nearly a *decade* before his anticipated retirement, "[choosing my successor] is the most important decision I'll make. It occupies a considerable amount of thought every day."

When Welch retired in 2001 and turned over the reins to Immelt, he could be confident that the enterprise was in good hands.

Incidentally, Welch's predecessor, Reginald Jones, had created a document entitled "A Road Map for CEO Succession" seven years before Welch himself took over Jones's position at GE. Jones spent two years creating a list of 96 possible successors, all of whom worked within GE. Like Welch after him, Jones didn't have to look outside the company to find qualified talent. Some of the final candidates went on to head such companies such as Rubbermaid, Apollo Computers and RCA. In fact, more GE alumni go on to become chief executives of American corporations than the alumni from any other company.

Bill Gates was able to feel confident about his successor in the late 1990s. Gates, who founded computer technology giant Microsoft with Paul Allen in 1975, found a strong Number Two when he hired Steve Ballmer in 1980. Ballmer headed several divisions at Microsoft, including "Operating Systems Development" and "Sales and Support." In 1998, when Gates realized his role in running the company was keeping him from the critical task of product development, he chose Ballmer to become Microsoft's new president; in 2000, Ballmer succeeded Gates as CEO. Gates didn't have to venture outside the company to find new leadership – Ballmer's years of experience and management at Microsoft made him the prime candidate. So prepared and confident in his ability to run Microsoft, Ballmer even reportedly said of Gates, "I'm not going to need him for anything. That's the principle."

And, just like Wild sometimes served as a bridge between Shackleton and his crewmen, so too can second-in-commands close the gap between the CEO and his workforce. Mike Sheard, who founded and ran executive recruitment company MGM in England, realized at the age of 64 that he did not share the same values as the 28-to-32-year-olds that worked for him. Concerned that this lack of compatibility would hinder his role as leader, Sheard sought a Number Two that could close the generational gap. Not only did he find that in Phil McDonald, who Sheard hired as MGM's managing director, but he also found in McDonald somebody that could succeed him in the future. "He could be better at this than I am," Sheard admitted.

As often was the case with Wild, a strong second-in-command can take care of operations when a leader must concentrate on something else. Shackleton couldn't be everywhere at once – most people can't – but with Wild he had an extra set of eyes and ears on the expedition, watching over the crew when Shackleton was busy with other critical tasks, like forging plans to get his men home safe. The same dynamic has helped many companies flourish over the years. Fashion designer Calvin Klein built his successful label with the help of Barry Schwartz, whose focus on the managerial operations of the label allowed Klein's vision to come to fruition. Guy Laliberte, the mastermind behind Cirque du Soleil, acknowledges he is less intrigued by the day-to-day operations than he is by his production's creative aspects. These "routine" responsibilities are delegated to Daniel Lamarre, who skillfully manages the business side of things. Lamarre says of the partnership: "I'm very lucky because we are so complimentary. What Guy loves to do, I don't and what I like to do, he doesn't." And according to Robert D. Walter, who once headed health-care distributor Cardinal Health Inc. by himself, he couldn't have stimulated sales and gained new acquisitions during the 1990s if he hadn't hired President and Chief Operating Officer John C. Kane. "I saw hiring a No. 2 as increasing our growth rate and lowering our risk -- and taking advantage of what I think I do best," he said. Walter managed acquisitions and general company strategies, while Kane focused on day-to-day operations. In all of these cases, a strong Number Two complimented the leader's role by handling tasks that the CEO didn't. A great Number Two doesn't always have to mirror your own interests – but they should be qualified and ready to help you reach your goal.

No matter how strong, smart, and compelling a leader you are, support from an equally talented No. 2 can only add to your success – or maintain it if you leave. Choosing a capable and assured assistant will enhance, rather than diminish, your stature as a leader.

So the next time you're tempted to hire or promote a lesser personality as your second-in-command, remember the tremendous advantages that a strong Number Two can bring to the table. In choosing yours, look for someone whose strengths reflect and complement your own – someone who won't be a "yes man" but whose loyalty you can trust. Such a person can support your efforts to lead your team, and can serve as a qualified substitute when you need to be away. It may not mean life or death, as in Shackleton's case, but it can have a huge impact on your success in running an effective team.

The Endurance crew

LESSON THREE

Can You Sing?
It's All About the Team

"It always looked as if he was picking them out with a pin but each time he got the right man." – *James Fisher, biographer*

In the business world, credentials often outweigh an applicant's personality during the hiring process. An impressive resume that lists Fortune 500 companies and an Ivy League school can make a candidate look very appealing. And let's face it: hiring a person with exceptional qualifications can protect you in the future. If your new employee turns out to be a disaster, no one can question your decision to hire someone who looked so good on paper.

Shackleton certainly took experience into account when he interviewed men for his expeditions. But he knew that experience wasn't everything. He once said that when it came to choosing his crew, "their science or seamanship weighed little against the kind of chaps they were." He had learned the hard way on the *Nimrod* expedition that a seafaring background didn't necessarily guarantee a qualified crewman.

Rupert England, the *Nimrod*'s captain, had made numerous voyages to Africa and Antarctica. But all his experience couldn't make up for an anxious, indecisive personality. When the *Nimrod* got stuck in the ice near the shore of Cape Royds, where the rest of the expedition crew awaited its arrival, he was afraid even to try to move the ship toward land. As Frank Wild put it at the time,

England had "entirely lost his nerve." After enduring two days of paralyzed inaction, a frustrated Shackleton had to order the captain to go at full speed – breaking an accepted rule of the sea, which states that *no one* undermines the captain's authority. England had to be dismissed from his duties. Shackleton later told England that he was "quite capable of undertaking any ordinary navigation job," but advised he should avoid "Arctic or Antarctic work."

When Shackleton set about finding a crew for his *Endurance* mission, he wanted to make sure to rule out another Rupert England. So he placed the following ad in the newspaper:

> "Notice: Men wanted for hazardous journey. Small wages. Bitter cold. Long months of complete darkness. Constant danger. Safe return doubtful. Honour and recognition in case of success."

Who would answer such an ad? Certainly not the cautious Rupert! Shackleton was seeking men who could look past the warnings and see an exciting challenge that they felt capable of handling. He knew he needed men who were enthusiastic and optimistic – traits that would translate into confident and capable seamen. And he needed men who felt strongly about the expedition – so strongly that they were willing to risk hardship and danger. He certainly did not want men who were just going along for the ride.

"The men selected must be qualified for the work," Shackleton wrote about hiring men for his *Nimrod* expedition, "and they must also have the special qualifications required to meet polar conditions. They must be able to live together in harmony for a long period of time without outside communication, and it must be remembered that the men whose desires lead them to the untrodden paths of the world have generally marked individuality."

His interviewing techniques were just as unorthodox as his advertising strategy. "Can you sing?" he asked Reginald James, who was applying to become the ship's physicist. James was taken aback, to say the least – but it was a question Shackleton frequently asked candidates. He wasn't looking for good voices, of course. He was looking for cheerful temperaments and good spirits, qualities that he felt were critical to team unity and success. He knew that singing would be an integral part of the expedition; it raised group morale and helped foster a comfortable work atmosphere.

He looked for a good sense of humor, too. When Shackleton asked the bespectacled candidate James Macklin what was wrong with his eyes, Macklin responded, "Many a wise face would look foolish without specs!" Shackleton laughed, and Macklin was hired as ship's doctor.

There was one more personality trait Shackleton considered essential: candidates had to be willing to get their hands dirty, regardless of rank. He could not afford prima donnas on his expedition. When Christopher Naisbitt, an elegant navy officer, applied for a job with Shackleton's *Quest* expedition, Shackleton worried that this gentleman might refuse to perform the menial jobs that were necessary to help the ship run smoothly. So he set Naisbitt to work on a series of tasks that included scrubbing a kitchen and peeling potatoes. The candidate never complained once, and Shackleton promptly hired him.

Shackleton's aptitude for team-building didn't halt once he had compiled his crew. During the voyage, there were critical moments when Shackleton needed to divide his men into productive groups. Just as the choice of a strong Number Two was important, because Shackleton couldn't be everywhere at once, it was also important to put together working teams that could function well without his direct supervision, and could operate on their own if needed.

A particularly good case in point was the way he divided his men between the three lifeboats when they embarked on their journey to Elephant Island. As was his usual practice, Shackleton kept some of the weaker crewmembers with him among the eleven men he chose for the *James Caird*. Nine of the strongest sailors were chosen to sail the *Dudley Docker*, since it would be their responsibility to also take care of the smallest and least efficient of the boats, the *Stancomb Willis*. Because of its weakness, the *Stancomb Willis* required strong sailors at its helm, so among its eight crew members were four hard-working seamen. Since Hubert Hudson, whom Shackleton chose as leader of that boat, was an excellent navigator but a less than powerful leader, Shackleton placed the vigorous Tom Crean at Hudson's side.

In addition to creating well-balanced groupings when "sub-teams" were called for, Shackleton was also masterful at fostering and reinforcing the crew's team unity as a whole. We've seen that one of his methods was to rotate menial chores and tasks among everyone, regardless of rank. Not only did this help to eliminate resentment or jealousy – it also meant that every man, sooner or later, worked side by side with every other man. As a result, crewmen who might not otherwise have socialized with one another got to know each other, and often became friends. Shackleton knew it was of utmost importance that they could all work together. And he knew that the more bonds of friendships were forged among the crew members, the stronger would be the feeling of unity and comradeship. For the men of the *Endurance* crew, it's possible that a powerful sense of mutual warmth and unity made the difference between disaster and survival.

* * *

It's more than likely that you will not ask candidates to sing or peel potatoes as part of your interviewing process. You can, however, heed the valuable idea behind Shackleton's method. It's a simple idea but a powerful one: *Don't underestimate the importance of personality and disposition for team success.*

Anthony C. Hsieh, CEO for Zappos, the number one seller of shoes on the Internet, understands the importance of team disposition all too well. The 1,500 people who work for him have been described as "perpetually chipper" employees – many of which refer to the workplace as "home" and their co-workers as "family." It's no surprise, then, that Zappos was placed on Fortune's list for "Best Companies to Work For" in 2008. This nurturing environment lends itself well to one of Zappos's most highly acclaimed strengths – great customer service. With the ability to be themselves ("Create fun and a little weirdness" is one of Hsieh's credos), Zappos employees are allowed to go the extra mile for their customers through actions such as sending customers flowers if they've had a bad day or bending company policy by accepting returns on worn shoes and sending a new pair (and thank-you note) in its place. It's no surprise, then, that 75 percent of purchases come from repeat customers – and that the company saw a 41 percent sales growth in 2008.

Just as Shackleton tested candidates for expeditions by measuring their ability to perform menial tasks, so too does Hsieh weed out unreceptive applicants during company training sessions. During a four-week training session, applicants must spend two weeks on the phones, beginning each day at 7 a.m. There are no sick days allowed, and sometimes the sessions land during the holiday rush. At some point during the session, Zappos offers jobseekers $2,000 not to take a job with them – no questions asked, no strings attached. The point of the offer is to identify which applicants seriously want to work with Zappos – and those that don't. If an applicant felt wary about continuing with the company, the monetary offer is a quick way to seal their departure. Candidates who feel motivated and excited to join Zappos will reject the short-term monetary gain. "We do our best to hire positive people," says Hsieh, "and put them in an environment where the positive thinking is reinforced." This business model has even caught the eye of other companies, and in the summer of 2009, Zappos began holding two-day, $4,000 seminars to teach others how they can mimic the spirit of its corporate culture.

Amazon founder Jeffrey Bezos insisted only the best be hired for his company when it developed in 1996. His sentiment was reflected by one of Amazon's directors, John Doerr, who stated: "In the world today, there's plenty of technology, plenty of entrepreneurs, plenty of money, plenty of venture capital. What's in short supply is great teams. Your biggest challenge will be building a great team."

One characteristic Bezos looked for in potential managers was confidence; he wanted people secure enough to "hire other great people." This requirement ensured that the bar would be continually raised for potential employees, guaranteeing a workforce of qualified and exceptionally talented individuals. Even when others encouraged Bezos to hire people who were capable of handling the work, Bezos stuck to his guns. He only wanted people on the A-list. This type of mindset should start at the top of an organization; a CEO who hires a strong No. 2 sets the example for all those in management positions to only hire the best.

Bezos's method also meant hiring people who had talents unrelated to their line of work. Candidates who were also athletes and musicians were viewed more favorably. Bezos argued "[w]hen you are working very hard and very long hours, you want to be around people who are interesting and fun to be with." Perhaps Bezos did not ask candidates if they could sing, but like Shackleton, he placed a high value on individuals who brought more to the table than qualifications. Job interviews consisted of questions about values and personal interests. Required writing samples also helped shed more light on a candidate; some applicants wrote short stories and poems, others wrote out business plans. Candidates who could communicate well were judged more favorably. By hiring well-rounded individuals, Amazon can be sure they have an interesting and dynamic work force that contributes to their continued success.

A highly credentialed team may look impressive, but it's not necessarily a team that will work well with one another, or with you. So look up from those resumes now and then, and take some time to explore what kind of people you are interviewing. How do they get along with others? How do they approach hard work and decision-making? Are they optimists? Do they know how to smile?

And once you have your team established, don't assume your work ends there. It's important to stay alert to the strengths, weaknesses and particularities of each of your team members, and to observe how they react to one another when working together. In addition, it's a good idea to take note of team interaction *outside* the office as well, as it can translate into how a team functions at work. Perhaps you will notice that certain individuals have lunch together, or that others share a friendship beyond the workplace. Likewise, you might see that some employees hardly interact with each other at all. Use these observations as a starting point to form teams for non-essential projects.

At Zappos, Hsieh encourages employees to spend 10 to 20 percent of their time with team members outside the office. Managers who take their teams out to dinner or on hikes report "improved communication, greater trust and budding friendships" afterwards, citing a 20 to 100 percent increase in team efficiency.

"We encourage our employees to hang out with each other outside the office. Then they can get to know each other on a more personal level than when you pass by someone in the hallway and say hi," said Hsieh.

Bill Watkins, CEO of Seagate Technologies, the leading independent maker of hard drives for data storage, flies 200 of his employees to New Zealand every year for a crash course in teamwork. The week-long event, called Eco Seagate, consists of outdoor activities like hiking, kayaking and racing. Forming teams, or "tribes," to compete in adventure racing forces Seagate employees to work together in strenuous situations. Watkins dreamed up the event as a way to break barriers, enhance confidence, and make staffers better team players. Watkins says, "It's not about us holding hands and singing "Kumbaya," it's about being the best storage solution company in the world and accepting no less, and our culture's got to help us do that. When I sit in a room with people, I can tell who's been to Eco Seagate."

Watkins, who spent time in the army, believes teamwork that includes emotional bonds is the biggest motivator, and Eco Seagate promotes this bond; the races often result in tears as much as celebratory hugs. While there is no way to measure the effect of Eco Seagate on company profits and success, studies have shown that companies with a positive work culture have more commercial success than those without one.

While there are no formal retreats at Clif Bar, the company goes on rock-climbing weekends, ski trips and volunteers for Habitat for Humanity together. Informal bike rides occur spontaneously. Whether they realize it or not, they are operating as a team by enjoying each other's company and sharing their time together. This translates into a workforce that respects each other and works better as a unit.

Like Watkins's implementation of Eco Seagate, Shackleton also built his teams around common goals – whether they were do-or-die missions or simply keeping up with ship maintenance. Whichever the case, Shackleton's men always knew what was expected of them. You will find that giving your teams a clear, shared objective can enhance their productivity and prevent members from moving in different directions. Anything you can do to provide a common focus for your teams will improve their chances of success.

Building a great team can be like putting together a jigsaw puzzle: if it's done right, every part will fit together with every other. Call it team-building or call it chemistry – it's vital for any successful group endeavor.

Wordie, Cheetham and Macklin scrubbing the floors

LESSON FOUR

Camaraderie at 20 Below Zero: Creating an Optimal Work Environment

> "Certainly a good deal of our cheerfulness is due to the order and routine which Sir E. establishes where he settles down. The regular daily task and matter-of-fact groove into which everything settles into inspires confidence in itself, and the leader's state of mind is naturally reflected in the whole party." – *Capt. Worsley*

Hiring a great team is a great start. The next step is creating an environment where they can work together comfortably and successfully. Shackleton understood the importance of creating and sustaining this kind of environment. It was critical on board the *Endurance* – and even more critical when they were stranded on the ice and trying to make it home alive.

A number of things go into making a good working environment. Perhaps first on the list are the simple necessities of life: food, equipment, supplies. Shackleton made sure that his crew was provided with abundant good food and the finest equipment. In many cases, he allowed crewmembers themselves to select the gear they needed, to ensure that each man would have the tools he needed to do his particular job.

Second, people thrive when they know what is expected of them. Shackleton wasted no time establishing routine and order when the *Endurance* set sail. He enforced a regular schedule for work and meals, and made his expectations clear. This gave the entire expedition a sense of security and stability, even in the precarious and unpredictable world of Antarctic seafaring.

When the ship became ice-bound – and even more dramatically, after it sank – the loss of their established routines could have been highly demoralizing to the crew. Shackleton understood the danger, and acted immediately to put new routines and tasks in place. He encouraged his crewmembers to engage in a variety of projects, all focused toward the goal of team survival. Reginald James and Hubert Hudson attempted to assemble a wireless receiver to obtain transmission from the Falkland Islands. Frank Worsley made a device that estimated the speed of the ice drift. Frank Hurley developed an efficient method to thaw frozen meat at exceptional speeds.

And when he assigned tasks, Shackleton always took his crewmen's personalities into consideration. As a result, each man was performing work he enjoyed and felt motivated about – factors that immensely improved both productivity and personal satisfaction. Feeling engaged and energized by work was sometimes the only thing that stood between the crew members and despair. Perhaps just as important, each man kept busy. As Orde-Lees noted, "At present I never seem to have time to do half the things I want to do and my only fear is that I may one day get ahead of my tasks and find myself in the unenviable position of having nothing to do but read, sleep and eat." When a man marooned on an Antarctic ice floe can testify that his only fear is boredom, you can be sure his leader is doing a good job.

Shackleton's ability to match man with task was uncanny. Take the example of Orde-Lees, who was preoccupied with the fear that the ship would run out of supplies and was constantly hoarding his food. For most leaders, the obvious response would be a strenuous attempt to persuade the man that supplies were plentiful and there was no need for paranoia. But Shackleton, with his usual insight, had a different idea: he put Orde-Lees in charge of the ship's supplies.

It was a brilliant move for several reasons. In one stroke, Shackleton transformed Orde-Lees' weakness into a strength, easing his worry and giving him a sense of pride in contributing to the overall welfare – and, last but not least, ensuring that the supplies thereafter were scrupulously managed.

Shackleton couldn't transform all weaknesses into strengths, of course. But he could, and did, make sure that weaknesses were never pointed out publicly. If he noticed one crewman shivering, he ordered hot tea to be served to all. If he detected a lapse in loyalty on the part of one man, he reminded them all of the rules. Knowing a man's weakness did not mean exposing it. On the contrary, his goal was to preserve each man's dignity.

And though Orde-Lees was initially resentful when ordered to do "dirty work" often reserved for lower-ranking seaman, he soon understood the logic of Shackleton's orders: "I must say I think scrubbing floors is not fair work for people brought up in refinement. On the other hand I think that under the present circumstances it has a desirable purpose as a disciplinary measure; it humbles one and knocks out of one any last remains of false pride that one may have left in one…"

While this arrangement meant humility to Orde-Lees, it meant a gesture of respect to the lower-ranked seamen. Further, Shackleton made sure that every crewmember received the same food rations – and, when camping on the ice floes, the same opportunities for sleeping comfort.

Shackleton not only treated his men equally, but individually. He made a point of forging a specific personal connection with each one. Possibly, he remembered the loneliness and isolation he himself felt during his first years at sea, and his difficulty relating to his foul-mouthed, hard-drinking crewmates. Most certainly he recalled Robert Scott's military demeanor on the *Discovery* expedition: cold, tough and reserved. "That was the big difference between [Shackleton] and Scott, " crewmember James said. "Scott was much too Navy."

In contrast, Shackleton took an active interest in what his crewmen were thinking and feeling. He often engaged them in informal or personal conversations, sometimes keeping them company during the lonely hours of a night watch. "When everyone else retired to bed," a crewmember on the *Nimrod* wrote, "the night watchman was never surprised when Shackleton joined him for a half-hour's chat or to smoke a cigarette in the small hours before himself turning in."

Shackleton's emphasis on individuality also allowed his men to design their own bunks. One crewman had painted a fireplace in his compartment; another man's space was so well-appointed that it earned the nickname "No. 1 Park Lane." A snug living and dining area the crew created on the *Endurance* was known by all as "The Ritz." Shackleton was wise enough to know that such playful expressions of personality could help men keep their sanity in difficult circumstances.

Shackleton also understood the value of play in a work environment. "I knew how important it was to keep the men cheerful," Shackleton would later write. He knew that when team members were able to relax and enjoy themselves together, they were apt to work together much more productively. And he understood that a dire situation doesn't mean sacrificing levity – in fact, finding small avenues for pleasure and enjoyment becomes even more important in grim circumstances.

As Shackleton later recalled, "there seemed always plenty to do in and about our prisoned ship. Runs with the dogs and vigorous games of hockey and soccer on the rough snow-covered floe kept all hands in good fettle."

While the crew was still aboard the *Endurance*, they held weekly concerts that included playing instruments and singing. Shackleton himself was right in the center of things – when they held a singing contest, in fact, he won it. And on another occasion he entertained his men by dancing the waltz. Later, during an especially horrific spell of bad weather when they were camping on the ice, Shackleton sent Hussey to visit each tent with his banjo.

Music wasn't the only diversion Shackleton encouraged. His crew sometimes performed skits, played cards, or watched picture slides from photographer Frank Hurley's around-the-world travels. At one moment of particularly inspired silliness, the men – including their leader – all decided to shave their heads at once.

Shackleton made sure that mealtimes were daily occasions for camaraderie. Everyone sat down for meals together; there was no separation between officers and seamen. No matter how arduous the day's labors had been, it was a time when they could come together, socialize and connect. Now and then, Shackleton would boost their spirits with extra rations – always a cause for celebration.

"[T]he depression occasioned by our surroundings and our precarious position could to some extent be alleviated by increasing the rations, at least until we were accustomed to our new mode of life," Shackleton surmised. He was right. The increase in rations "soon neutralized any tendency to downheartedness." As one crewman wrote, "Meals are invariably taken very seriously, and little talking is done till the hoosh is finished."

Holidays such as Christmas were enthusiastically celebrated, and birthdays were honored with gifts, singing and photographs. All of these rituals and celebrations gave the crew a comforting sense of normalcy in a situation that was anything but normal.

While Shackleton was gifted at providing his men with opportunities for fun and enjoyment, he was equally determined to make sure things never got out of hand. Alcohol was limited to rare occasions – and when it was served, it was never in large quantities. And while sports and games were encouraged, they were stopped if there was a risk of injury. The men were keenly aware of this. When Timothy McCarthy got hurt in a hockey game, he managed to get stitches without Shackleton finding out about it – which would have meant an end to intramural hockey.

* * *

How can you create a productive working environment for your employees? Chances are, they are well fed and abundantly supplied. You might take a leaf from Shackleton's book by allowing your employees some say about the tools they will use. You can't be an expert in every area of software and technology – so your employees in each of those areas will probably make more informed choices than you could about what they need in order to do their best work.

Stability, order, and clarity of expectations are every bit as important in a corporate office as they were on Shackleton's boat.

Having their time and their work sensibly and clearly structured will make your team members more creative, not less. A stable environment provides a firm basis for initiative, imagination, and productivity. And it's up to you to make sure that no one is doing busywork. If Shackleton could give every man on the ice floe meaningful work, you can do it for every cubicle dweller.

Further, by taking an active interest in each individual, you can build personal connections with your employees that encourage trust, respect and amity. Taking a personal interest also allows you to access your employee's strengths and weaknesses, permitting you to delegate tasks accordingly. Keep in mind Shackleton's neat trick of turning a weakness into a strength. Experiment, from time to time, with putting a procrastinator in charge of the scheduling, or asking a daydreamer to solve a problem by thinking outside the box. Sometimes a true leader is the one who can look at liabilities and imagine assets.

The appreciation of the individual also includes respect and equal treatment for all employees. You might not want to ask your higher-level employees to make copies or sort the mail, but it's important to avoid preferential treatment. Treating certain employees with more respect and perks than others will inevitably result in resentment and frustration for those excluded, leading to low motivation and poor productivity. Every employee needs to feel just as valued and respected as every other. By the same token, according all employees the same level of respect will go a long way toward creating an upbeat and productive working atmosphere.

And don't forget to play! In today's corporate world, competitive pressures and stresses sometimes tend to make for a fairly relentless work environment. And certainly success requires that employees approach their jobs in a serious and focused way. But we can learn from Shackleton's example that sometimes people work better together if they have a chance to enjoy themselves and each other.

Google co-founders Sergey Brin and Larry Page understood that an ideal work environment was key to productivity. "We think a lot about how to maintain our culture and the fun elements," said Page in 2004, when the company went public. "We spent a lot of time getting our offices right."

Indeed, at Googleplex, the company's headquarters in Mountain View, California, Brin and Page introduced a number of measures that ensured workers would look forward to coming to work and be content once they got there. They hired a chef to serve free, wholesome meals to all employees. It seems like a simple perk, but it had important ramifications: employees stayed close to their desks instead of venturing off company grounds for a meal; less time was wasted on employees traveling to get lunch; healthy food options prevented unhealthy eating habits, which could reduce productivity (not eating or eating poorly has been shown to affect concentration and energy levels on the job); and employees spent more time bonding together. In a Google employee survey, nine out of ten people cited the food as what they liked best about their jobs.

And with such credos as "you can be serious without a suit" and "work should be challenging and the challenge should be Fun," Brin and Page redefined corporate culture. Similar to the crewmen's personalized bunks on the *Endurance*, Google employees are encouraged to individualize their workspace. Some cubicles are wrapped in aluminum foil, others pay tribute to Star Wars or other pop culture entities. One cubicle was even described by a Google employee as a "Zen oasis" – equipped with mood lighting, bean bags and small bubbling water fountains.

Foosball, pool tables, volleyball courts and assorted video games can be found at most Google offices, and informal groups participate together in meditation classes, film clubs and salsa dance clubs. Not only does this allow employees to relax and blow off steam, it forges bonds between co-workers that can strengthen the team dynamic. Of course, not all companies can offer such social spaces, but playtime doesn't need to be extravagant. Organizing low-key social events around employee's interests or having informal gatherings can encourage social bonding. If you're concerned about "too much" play, recreational times can always be limited to specific parts of the day or week, or activities and events can be designated for after work hours only.

Onsite doctors and laundry facilities are also readily available. Such an environment encourages employees to come into work early (for free breakfast until 8:30 a.m.!) and stay in the evening. It also keeps employees happy, comfortable and loyal – all components necessary for effective and innovative productivity. It's no surprise, then, that Google was placed at the top of Forbes' Best Places to Work list in 2007 and 2008. According to chief culture officer Stacy Sullivan, Google's corporate culture is "one of our most valuable assets."

Erickson, head of Clif Bar, took the initiative to re-introduce playfulness at the office when he sensed people were getting too serious. The company holds cooking contests modeled after TV's "Iron Chef" show, throws annual "Martini and Weenie" parties, and goes on yearly camping trips. Every day is "Bring Your Dog to Work Day," which adds a dose of canine fun to the workplace. As one employee said, "We don't take ourselves too seriously. We have fun and our advertising shows it. The creativity that we put forth shows that we know 'it's only food' and are humble about it."

Other companies have also instituted environments that rely less on the traditional corporate culture, and more on the comfort and well-being of its employees. David Olson, President and CEO of Patagonia, the outdoor apparel company based in Ventura, Calif., believes his company "cannot build great quality products without a great quality work environment. If you overlook any piece in the puzzle, there is a good chance you'll miss it all."

So what are some of the elements of Patagonia's work environment? Olsen allows his workers to dress as they wish, even if that means bare-footed. Employees are encouraged to take a surfing break at lunchtime or even play volleyball. Patagonia was also one of the first companies to open an on-site child care center. Children playing in the yard and having lunch with their parents helps keep the atmosphere less corporate. The company provides child-care subsidies, buses to transport children from local elementary schools to the company's offices, and parent education seminars during lunch hours. Two months paid maternity leave is also provided for both mother and father after the birth or adoption of a child.

Rather than purposely breaking from the typical work environment, Patagonia simply believes their work culture upholds their own traditions and values. These values, in turn, translate into superior production. In 2006, the company amassed over $275 million in sales revenues, and was even praised by President Clinton in 1996 as an "impressive" example of corporate responsibility. And while employee turnover is 20 percent industry-wide, at Patagonia the turnover is a slim 4.5 percent.

Similarly, Costco Wholesale's turnover rate is 20 percent compared to the industry average of 65 percent. The low turnover, as well as a low theft rate, can be attributed to the positive work environment instituted by co-founder and CEO Jim Sinegal. Costco employees enjoy the highest pay rates in the industry (while a Wal-Mart employee averages $9.68 an hour, Costco employees earn $16; they also earn more on Sundays). Generous health benefits are extended to those who work part-time. And though only 18 percent of the workforce is unionized, Costco extends most of the benefits outlined in union contracts to its non-union workforce.

When asked about Costco's generosity to its employees, Sinegal tells *Fast Company* magazine, "We owe that to our employees, that they can count on us for security. We have 140,000 employees and their families; that's a significant number of people who count on us."

As Chief Financial Officer Richard Galanti explained to *Working Life*, "From day one, we've run the company with the philosophy that if we pay better than average, provide a salary people can live on, have a positive environment and good benefits, we'll be able to hire better people, they'll stay longer and be more efficient."

At the women's clothing giant Eileen Fisher, employees are granted "wellness" and education allowances that total $2,000 per year. The money can be spent on massages, yoga, gym dues or other healthy activities. The education allowance to develop new skills like dance or guitar helps employees grow as individuals.

"I think it makes people feel good about themselves," Fisher, who started the company in 1984, said. "They know we care about them as people and as whole people, not just what they can do and produce and - you know, bring to the company."

"We do it not only because it feels right, but we also know that we then get the best of people," says EF's Chief Culture Officer Susan Schor. And believing that the well-being and development of individuals improves the company dynamic overall, employees are encouraged to seek new challenges for themselves and other team members. "This is a company where you can start out doing one job function and, if you have a passion or an interest in something else you can wind up doing something totally different just because you wanted to, and because there is the space here for you to do that," says EF's Director of Architecture Peter Scavuzzo.

And Fisher believes the energy and money spent on her employees only helps the company's success and longevity. "You know, if you're paying attention to what you care about and what you love – and for me, how the whole thing comes together — then it tends to work at the bottom line," Fisher said. "Bottom line is really just numbers that reflect what's happening in the center. And so you pay attention to what's happening in the center, and when that's right, the numbers follow."

Indeed, there are no signs of Eileen Fisher slowing down anytime soon. The company has been growing five to ten percent annually over the past five years, and amasses a quarter-billion a year in sales.

Understanding the connection between happy, well-treated employees and the group dynamic is critical in the development of an optimal work environment. This doesn't mean that offices should exist without rules or boundaries, but when employees have time to relax and be themselves, you can get the best out of each individual.

Wild standing next to the shipwreck

LESSON FIVE

Sailing Uncharted Waters: Adapt and Innovate

"A man must shape himself to a new mark directly the old one goes to ground." – Shackleton

The true test of a leader occurs when something goes wrong. And sooner or later, something always goes wrong. No matter how carefully you've planned, you will eventually be faced with something you didn't prepare for.

On the *Endurance* expedition, more things went wrong than anyone had ever anticipated. Time after time Shackleton and his men would overcome a problem only to confront another and more drastic one. And it was here, in his manner of dealing with setback after setback, that his genius for leadership really stood out.

One element of this genius was his ability to think ahead and envision alternative scenarios. While his crew waited out the long winter on their ice-bound ship, Shackleton was anticipating the possible need to abandon it. He kept this thought to himself but undertook quiet preparations, moving some food stores to the deck of the boat where they could be quickly transported. This made the move easier when the time finally came.

His diary reveals that the order to abandon ship was not easy for him to give. "It is hard to write what I feel," he wrote. "To a sailor his ship is more than a floating home…Now, straining and groaning, her timbers cracking and her wounds gaping, she is slowly giving up her sentient life at the very outset of her career." He showed no trace of his sadness to his crew, however, focusing instead on the urgent tasks that lay ahead. Macklin wrote that Shackleton "did not rage at all, or show outwardly the slightest sign of disappointment; he told us simply and calmly that we must winter in the Pack [ice]; explained its dangers and possibilities; never lost his optimism, and prepared for the Winter." Shackleton's demeanor was equally calm when the *Endurance* sank at last into the sea. Turning to his men, he said simply, "Ship and stores have gone, so now we'll go home."

Shackleton knew instinctively that every leader sets the emotional tone for his team. If the leader expresses despair or regret, the team will inevitably lose heart. But if the leader communicates confidence, his optimism, too, will be infectious. The calm, positive spirit Shackleton showed his men helped them face what lay ahead.

And what lay ahead, of course, was a series of steadily mounting difficulties. Shackleton's way of facing these challenges exemplifies another aspect of great leadership under duress: the ability to adjust. When the crew first abandoned ship, for example, Shackleton weighed a number of factors and decided the best course was to walk toward land, dragging the lifeboats. Within a day, the perils of this plan were obvious. The snow was nearly impassable, the men were already exhausted, and the lifeboats were in danger of being wrecked. Shackleton wasted no time in changing his mind. He reassessed the situation and made a new plan for the crew to set up a temporary camp on the floes.

Later, when the ice floes melted and the crew finally boarded their lifeboats, Shackleton changed his mind over and over about which direction to sail – four times, to be exact, over a four-day period. His decision to head towards Elephant Island became final only when he noticed how "seriously worn and strained" the men looked, noting their "lips were cracked and their eyes and eyelids showed red in their salt-encrusted faces." In a lesser leader, such rapid changes of plan might have aroused misgivings in the ranks. But as James noted, the crew always trusted Shackleton's instincts. "Well-settled plans would suddenly be changed with little warning and a new set made. This was apt to be a little bewildering but it generally turned out for the good. This adaptability was one of his strong points."

It was crucial that Shackleton never brooded over missed opportunities or failed strategies. Had he concerned himself with the past, he would have been unable to move forward. "As always with him," wrote Macklin, "what had happened had happened; it was in the past and he looked to the future."

Shackleton's ability to adapt to changing circumstances also reflects his innovative spirit; his mind was always wheeling with new ideas. His method of financing the *Endurance* was a very novel approach. Instead of solely relying on government contributions, Shackleton licensed the rights to photographs, film and stories compiled during the journey. In fact, all the crewmen on the expedition were required to give their diaries to Shackleton at the end of the trip (which is why many men kept a second, private, diary). He also made important contributions to the world of seafaring. He introduced improvements to explorer Fridtjof Nansen's fur apparel by changing the heavy, bulky material to a lighter windproof suit. And, unlike other explorers who packed equipment in whatever container they could find, Shackleton used 2,500 uniform Venesta packing cases on his *Nimrod* expedition, making packing easier and using those same cases as furnishings on the boat. Not content to merely mimic what other explorers had done before him, Shackleton's innovations improved the quality of his expeditions – and his sharp mind kept his men afloat even when disaster struck.

* * *

It's to be hoped that as a leader, you will never face any misfortune as drastic as your ship sinking beneath the sea or the ice floe on which you are stranded drifting farther and farther from land. But when a project important to you flounders, it can feel like disaster.

What can we learn from Shackleton about leadership in this situation? First, it's important to be thinking ahead whenever possible, anticipating alternative scenarios and preparing for them. Second, when a decision you've made proves to be leading in the wrong direction, don't be afraid to change course. Sometimes we're tempted to stick doggedly to a position we've taken, no matter what the outcome. But the leader who can show flexibility is far more likely to overcome challenges and difficulties than the one who is committed to staying on a mistaken course.

And being able to adapt well also calls for an innovative spirit. It's one thing to change plans when necessary, but if the alternate strategies are stale and uninspired, you'll go nowhere. Many lagging companies have been revitalized when new ideas catapult them to new heights.

In the early 1990s, Apple began to suffer after the release of a number of failed consumer-targeted products. Subsequently, the company's market share and stock value continued to slide. Apple realized that the Macintosh platform was becoming outdated and, under the leadership of Steve Jobs, concentrated on creating products that were both functional and elegant – two factors that play a huge role in a product's public appeal. With the release of the iMac in 1998 and the supremely successful iPod and iTunes store in 2001, Apple found itself back on top once again. "With iPod, listening to music will never be the same again," Jobs declared. Customers agreed; over 100 million units of the iPod were sold six years after its release.

The makers of Cranium, one of the best-selling board games since Pictionary debuted in 1985, used both innovation and adaptation to bring their product to the market. The creation of Cranium itself was an innovative idea. Richard Tait and Whit Alexander developed the game after Tait and his wife played board games one night with some friends; while they lost miserably at Scrabble, they flourished playing Pictionary. Like most games, Scrabble and Pictionary focus on one specific skill set (i.e. trivia, drawing, words). What about a game, Tait thought to himself, that incorporated all sorts of skills so that everyone could have a shining moment? Enlisting the partnership of Whit Alexander, Cranium was born in 1997.

But Cranium did not follow the same path as other board games before it. Having missed its opportunity for selling the game to big distributors for winter distribution (Toys "R" Us, Wal-Mart), Tait and Alexander needed to reshape their approach. They decided to make a market of their own, placing Cranium in stores where the game would stand out instead of getting lost among a sea of other games. While brainstorming at a local Starbucks, the men realized the customers in line represented their target audience: smart, affluent 25-to-35-year olds. After several meetings with Starbucks executives, the coffee giant accepted the game; by November 1998, Cranium appeared in 1,500 Starbucks nationwide. Tait and Alexander soon approached Barnes & Noble – which didn't carry board games – about stocking Cranium. After employees played the game, Cranium was accepted into Barnes & Noble stores.

But Tait and Alexander still had to publicize their game. Television advertising, the go-to venue for the toy industry to reach millions of people, was far too expensive. So instead, the men turned to an alternative PR approach. They recruited 110 radio stations around the country to have their DJs read Cranium questions over the air and award games to callers with the correct answer. "Hearing a clever question or two gave a mini-trial to a broad audience," says Tait. Game-playing events were also held at Starbucks locations. Sales of the game snowballed. By early February 2001, Cranium had amassed $18 million in sales. By early 2002, one million copies of Cranium had been sold, and the game had become the toy industry's first huge hit since Pictionary in 1985 and one of the best selling games for that year. In an industry where 50 percent of games fail in their first year, and of the remaining 50 percent, half fail in their second year, Cranium is a golden exception. Had Tait and Alexander followed the beaten path of board games before them – or if Jobs led Apple to continually regurgitate the same product – both companies might not be experiencing their continued growth and success.

Throughout its history, leaders of Patagonia have used innovation to bring improved products to the outdoor apparel and accessories industry. By 1970, the company (known then as Chouinard Equipment) had become the largest supplier of rock-climbing tools in the United States. They were also blasted for the damage their supplies were causing to rocks. The company could have continued selling their products to rock-climbers with a dark cloud hanging over its head, but instead founder Yvon Chouinard re-dedicated the company to the environment and created pitons that wouldn't damage rocks: aluminum chocks that could be wedged by hand rather than hammered in and out of cracks. The chocks debuted in Chouinard's catalog in 1972 and sold faster than they could be made. Chouinard didn't have to compromise his dedication to the environment. Instead he adapted his products to serve the environment – and those that love it – better.

The company also introduced brightly colored men's apparel to the world of active sportswear (until the late 1960s, men wore gray, drab colors outdoors). While the mountaineering community still relied on moisture-absorbing layers of fabric, Chouinard introduced insulating layers of clothing that did not absorb moisture, inspired by the apparel of North Atlantic fisherman. In 1980, the company introduced the idea of proper layering to the outdoor community, debuting an insulating long underwear made of polypropylene that didn't absorb water. The company was clearly on the cutting edge.

With its innovation came rapid success and growth. The brightly colored apparel – meant for outdoor types – became a fad among fashion consumers. Inc Magazine placed Patagonia (as it was now known) on the list of the fastest-growing privately owned companies. This halted in 1991, when a recession deflated sales. Banks were calling back their loans to the company. Patagonia had to cut costs, dump inventory and lay off 20 percent of their workforce. It could have been close to the end for the once thriving company.

But this downturn caused Patagonia executives to reevaluate their game plan. Much like Shackleton – who reassessed plans of action as the situation dictated – Patagonia curbed their growth and borrowing to avoid future trouble. Concentrating on apparel and accessories for the outdoor community remains its core business, and the company isn't looking to expand its base.

Of course, some companies maintain a level of innovation before setbacks even occur, continually advancing their technologies and improving their product. One company continually touted for its innovative spirit is the package delivery company, FedEx.

"Looking back on 25 years at the forefront of a dynamic and evolving industry," said founder and current CEO Frederick W. Smith, "two things have remained constant at FedEx: change, and the ability and willingness of our employees to embrace change on behalf of our customers."

Indeed, ever since Smith famously stated in 1978 that "the information about the package is just as important as the package itself," FedEx has thrived by implementing cutting edge technologies that give customers access to almost real-time information on the status and location of packages delivered worldwide. Some firsts? Pioneering the use of wireless technology for shipping with the introduction of the Digital Assisted Dispatch System (DADS); installing computers in delivery vehicles for tracking capabilities; offering shipping and package-status tracking via the company's website; offering delivery for 10:30 a.m.; and aligning with the U.S. Postal Service to place FedEx Drop Boxes at post offices nationwide.

Though FedEx has revolutionized the delivery industry, it remains committed to the development of even newer technologies. The company kick-started the FedEx Innovation Labs, an environment dedicated to collaborative thinking about critical technologies, such as advancements in scanning, robotics, pervasive computing, social networking and more. This supplements an already existing research & development department. In addition, the company runs the FedEx Institute of Technology, situated at the University of Memphis. A versatile, high-tech facility, the Institute is ideal for think tank sessions, corporate retreats, training and national conferences.

FedEx management instills a culture of innovation throughout the entire company, not just those working in IT. "People are engaged and empowered to come up with ideas and present those ideas," Robert Carter, FedEx's executive vice-president, told *Business Management* magazine. "We believe that everybody has a responsibility to innovate and think of the next best thing."

Smith echoes Carter's sentiment. "It is the FedEx people – the entire team," Smith says, "who are responsible for the company's phenomenal success."

FedEx also keeps a pulse on its customers' needs and wants to help ignite innovation. After all, who knows what a customer needs best than the customer itself? Even though satisfying customer demands may increase costs for the company, it pays off in the long run. Take, for example, FedEx's overnight delivery service. Express service for delivery by 10:30 a.m. the next day was a popular shipping option among FedEx customers, and the use of this service was increasing. However, customers also wanted an option that didn't cost as much but still provided overnight delivery. So, FedEx introduced Standard Overnight delivery that would guarantee a package would arrive by the next afternoon instead. Many customers chose this cheaper alternative, thus giving less of their money to FedEx. But the service was so popular that the increased customer and package volume offset any lost revenue.

By continually trying to create new technologies and advancements, FedEx retains itself as an industry powerhouse. Most significant is the idea that all FedEx employees (and customers) can help jumpstart new ideas – not just those with certain qualifications and degrees.

The John Deere Company – which manufactures, distributes, and finances a full line of equipment for use in agriculture, construction, forestry, and lawn and grounds care – mirrored FedEx's innovative spirit by developing new products and technology even during economic and industry downturns. When the agriculture industry was sputtering in the 1980s, former CEO Robert W. Lane pushed for change and new products as the best path to revenue stability. Under Lane's leadership, John Deere inserted more power and functionality into its tractors and combines. It was during this decade that the company developed and manufactured the John Deere Gator, an award-winning alternative to all-terrain vehicles (ATVs). Though the powerful vehicle initially appealed to the military and pro-football teams as an efficient way to move equipment and injured persons, sales exploded when average customers realized how useful the vehicle could be in various situations.

Following its success with the Gator, John Deere implemented new improvements to the vehicle to increase its versatility, even creating the "TurfGator" for use on golf courses. The Gator transformed from a specialty vehicle to a product popular amongst a wide range of people.

Lane was firm in his commitment to change: "We should not just be looking to do things a little bit differently but be prepared to change completely the way we do business," he told the *Financial Times* in 1999. During another economic downturn in 2001, while other companies delayed pursuing product development, John Deere introduced 50 products into the agricultural market.

"I don't remember a time when Deere delayed the next generation of a product," recalls John Lawson, who worked at Deere for 44 years.

Even if profits were dropping amid tough economic times, John Deere maintained its commitment to deliver the best product and strengthen its competitiveness. Imagine if Lane decided to refrain from developing the Gator during the 1980's slump? While economic downturns can be a tricky time for any company to launch new – especially expensive – products, this doesn't mean that inventive ideas must come to a halt. Keeping the creative juices flowing can keep you one step ahead of the competition.

Staying inventive can help retain the success of your company. But when you change direction – whether intentionally or due to external factors – there's no point in ruing your mistakes. Shackleton got his men home safely because he was always looking forward, never wasting energy on "should've" or "could've." If your company finds itself in a funk, don't dwell on the past – it will only hold you back from discovering the innovative ideas and plans that can propel you into the next stage of your success. As Shackleton himself once said, "In trouble, danger, and disappointment never give up hope… The worst can always be got over."

Hurley and Shackleton sit outside their tent in Patience Camp

LESSON SIX

Be my Tent Mate: Keep Dissidents Close

"Put footstep of courage into stirrup of patience." - *Shackleton*

Everyone knows that when picking a roommate or confidante, it's important to choose someone you trust and get along with. Right?

Not necessarily...not if you're Shackleton. Consider this: among his crew members, carefully chosen as they were, there were inevitably a few naysayers. There was photographer Frank Hurley, who was sometimes heard making critical remarks about Shackleton to other crewmen. There was navigating officer Hubert Hudson, an argumentative individual prone to disagreements. And there was physicist Reginald James, whose constant anxiety could be counterproductive.

It would have been easy for Shackleton to keep the negative crewmen at a distance. Instead, when setting up camp on the ice, it was precisely these fellows he picked to be his tent mates.

"He collected with him the ones he thought wouldn't mix with the others," Greenstreet wrote. "They were not so easy to get on with, the ones he had in his tent with him – they were quite a mixed bag."

Living in close quarters with the three men, Shackleton was able to win them over one by one. He treated Hurley like a confidante, which flattered the dissenter's ego and inspired in him a newfound respect for the expedition leader. Hurley developed a "great admiration for the boss – who is ever considerate & kindly disposed – an excellent comrade." Shackleton managed to divert Hudson's argumentative personality away from potentially destructive quarrels and toward lively debates over innocuous subjects. And James' nervousness about the dangers of the expedition was soothed by his proximity to the expedition leader.

Shackleton was equally clever months later, when selecting which crewmen would join him on his voyage aboard the *James Caird* lifeboat from Elephant Island to South Georgia. He knew it was crucial that the men who remained on Elephant Island stay optimistic and productive while they waited for help. Negative or overanxious crewmembers could destroy morale while Shackleton was away, with possibly disastrous consequences.

In this regard, he was particularly concerned about carpenter Harry McNeish and seaman John Vincent. Vincent was known by his crewmates as a bully who sometimes used his robust body frame to intimidate them. Though his demeanor had improved (with guidance from Shackleton) during the course of the journey, he could not be trusted to remain peaceable if left behind on Elephant Island. McNeish, while not a bully, sometimes seemed to feel superior to Shackleton and critical of his decisions. When he refused to obey Shackleton after the *Endurance* sank, the two men nearly came to blows. Shackleton, in a characteristic move, reminded the entire crew of their obedience by reading the Ship's Articles out loud. It stated, in part, that all crewmen must "be obedient to the lawful commands of the said Master...whether on board, in boats or on shore." But in his diary, he confessed, "I shall never forget [McNeish] in this time of strain & stress." Not only was McNeish's loyalty questionable, but he often expressed pessimism about the crew's chances of ever reaching land safely.

But Shackleton was acutely aware that if left behind on Elephant Island, Vincent and McNeish could cause dissention and negativity to spread among the entire crew. He also recognized that they were both skilled additions to the boat journey; Vincent was a good sailor and McNeish, a skilled carpenter. So along they came with Shackleton.

It's interesting to note that Shackleton's gift for diplomacy had showed itself long before the *Endurance* expedition, in a context where he was not leader but team member. Antarctic explorer Robert Scott accused Shackleton publicly of causing the premature return of his 1901 *Discovery* expedition; years later, when Shackleton's *Nimrod* expedition came closer to the South Pole than anyone had before, Scott spread rumors that Shackleton had fabricated how close he had come to the Pole. Shackleton felt hurt and betrayed by these false accusations. But he never criticized Scott in public, and he made a point of being civil whenever they met. In 1910, Shackleton even delayed the *Endurance* project in order to help Scott organize his *Terra Nova* expedition.

Shackleton's diplomatic attitude proved wiser than he knew. When Scott died on the return journey of the *Terra Nova* he was mourned as a national hero. Had Shackleton developed a reputation as Scott's adversary, he would have lost valuable public support.

* * *

In corporate life as in Antarctic exploration, every team has its share of naysayers and critics. For a leader, it's naturally tempting to push these malcontents as far away as possible, so as not to have to deal with them. How many corporate executives can you think of that cultivate close relationships with their most outspoken antagonists?

Shackleton understood that, paradoxically, the very safest place for a problematic team member was close to him. The same principle holds true for corporate team-building. If you keep your critics at arm's length, you won't know what they're saying or doing – and you'll give them the leeway to do real damage. They're a lot less likely to cause harm if they remain in your view…and perhaps even in your tent.

Also, some companies try to avoid dissidents because it implies that they are not in line with the mission of the company. More appropriately, we should look at the value of employees who offer a wide range of viewpoints, including those who may vary from the majority. Those dissenting opinions are not always easy to come by as they are often shunned, however when companies embrace dissenting opinions they gain an invaluable asset. Having a diverse management team that values honesty and contrary thoughts and ideas can prevent companies from making devastating mistakes, can steer a project in a new direction, and can create possibilities for innovation.

A great historical example of this can be found outside the corporate world – within Abraham Lincoln's Presidential Cabinet in 1861. Lincoln ran against three men in the Republican primary: Senator William Henry Seward; Ohio Governor Salmon P. Chase and judge Edward Bates – three men whose political backgrounds were far more extensive than Lincoln's. When Lincoln, virtually unknown before 1858, won the primary, they were shocked. And they were even more stunned when he won the presidency. Each man couldn't believe that a "prairie lawyer" from Springfield was now running the country.

But, in Shackleton-esque fashion, Lincoln chose all these men to serve on his Cabinet. Seward was appointed Secretary of State; Chase became Secretary of the Treasurer; and Bates became the U.S. Attorney General. When later asked why he chose these men, Lincoln replied, "We needed the strongest men of the party in the cabinet. These were the very strongest men. Then I had no right to deprive the country of their services." Instead of pushing away three men who not only wanted his job but thought they would be better at it, Lincoln made the unprecedented decision to incorporate them into his political family.

The early days of Lincoln's presidency weren't easy. Seward initially believed, like many others, that he would be the "premier" of the new administration, and undermined Lincoln's executive leadership. He even penned a memorandum accusing Lincoln of being "without a policy either domestic or foreign." Lincoln did not terminate Seward; he knew such a shake up might cause even bigger problems. Instead, he spoke personally with Seward and asserted his own policies and the necessity to carry them out himself.

Lincoln's approach worked. He soon won Seward over, and the two became very close. They spent time together outside of the

White House, exchanging stories of their personal and political past, and Seward became one of the president's most loyal and valuable supporters. He commended Lincoln as "the best and wisest man he [had] ever known."

Imagine how the story would have turned out if Lincoln initially ousted Seward from his Cabinet? The powerful Senator might have used his influence to undermine Lincoln's capacity as president, potentially causing a political backlash against Lincoln during such a turbulent time in U.S. history. Instead, Seward became one of Lincoln's strongest allies.

Attorney General Bates was eventually won over by Lincoln as well. Much like Seward, Bates initially underrated Lincoln's capacity to be president. Lincoln's repeated assertion of his leadership forced his team to overcome pettiness and get behind his policies.

Chase, however, remained a constant thorn in Lincoln's side. Lincoln valued Chase's ability as Secretary of the Treasurer, and squashed any rebellion by re-asserting his authority much like he did with Seward and Gates. Chase began a game of threatening resignation to force Lincoln's hand on particular policies, but Lincoln would never grant his dismissal. It was only after Lincoln was safely re-elected into a second term that he – to Chase's shock – agreed to relieve Chase of his duties. But instead of casting Chase off, Lincoln appointed Chase as chief justice of the United States, preventing any political backlash over Chase's discharge.

So next time you get wind of an employee grousing about your leadership, consider taking the person to lunch. Better yet, install the person in the office next to yours. Proximity can be your best strategy – just think of Shackleton in that lifeboat.

The crew attempts to cut through the ice

LESSON SEVEN

Breaking the Ice: Communicate

> "[H]e would get into conversation and talk to you in an intimate sort of way, asking you little things about yourself – how you were getting on, how you liked it, what particular side of work you were enjoying most – all that sort of thing…This communicativeness in Shackleton was one of the things his men valued in him; it was also, of course, a most effective way of establishing good relations with a very mixed company." – *Alexander Macklin*

When the temperature is 50 degrees below zero, you might think communication is beside the point. On the contrary. The more extreme their situation, the more urgently team members need their leader to speak to them honestly and directly. Shackleton knew that if he wanted the best from his men, he had to communicate with them.

When he made the decision to abandon ship, for example, he gathered his men around him. He was composed and articulate. He explained their circumstances and the reasons for his decision. Looking forward, as always, he announced the new plan of action: they would set up camp on the ice and then embark on a march northwest towards Snow Hill, where emergency stores had been left by a previous expedition. Since the ice floe they were on was moving slowly north, they would march west; the drift of the ice would help move them in the direction of Snow Hill.

He did not try to sugarcoat their situation. By the same token, he communicated calm resolution and confidence in his plan. And he ended by expressing gratitude for their hard work.

"It must have been a moment of bitter disappointment to Shackleton," Macklin wrote, "he lost his ship, and with her any chance of crossing the Antarctic Continent, but he shewed it neither in word or manner. As always with him what had happened had happened: it was in the past and he looked to the future."

"It was a characteristic speech," wrote Hussey, the expedition's meteorologist. "Simple, moving, optimistic and highly effective." The men responded with heartfelt expressions of loyalty.

But Shackleton also understood the importance of timing when communicating his thoughts and plans. In many dire situations, like when it came time to abandon ship, Shackleton's blunt, honest speech was necessary. At other times, Shackleton chose to delay vocalizing his plans for the sake of group morale. This was the case when the men finally made it ashore Elephant Island after a tumultuous time at sea. Surveying their landing spot on the island, Shackleton observed evidence that waves could crash onto their resting site due to strong gale winds, wreaking havoc and irreversible damage to the boats; he knew they needed to set sail again and locate a better spot along the coast of the island.

But he also understood the men were exhausted and overjoyed to have finally reached land: "I decided not to share with the men the knowledge of the uncertainties of our situation until they had enjoyed the full sweetness of rest untroubled by the thought that at any minute they might be called to face peril again. The threat of the sea had been our portion during many, many days, and a respite meant much to weary bodies and jaded minds." Early the next day, Shackleton sent Wild to find a more suitable spot for the men to camp, and the (now rested) crew sailed in shallow waters to their new destination.

Shackleton knew that speeches were only half of the communication process. The other half, listening, was equally important. So while he did not hesitate to make final decisions on important matters, he always reached out to his crewmen for their opinions. He listened to Worsley's insistence that all three lifeboats should be dragged along on the march. He listened to Greenstreet's suggestion that the crew stockpile food as a precaution. Ultimately he did not follow either of these suggestions – but he heard them. Asking for his men's ideas and thoughts was his way of making sure they felt included and valued – and, often, it helped him to formulate his own plan of action.

Shackleton was a master of more subtle approaches to communication as well, when circumstances called for a bit of indirection. If he knew that a pending decision was likely to be unwelcome in certain quarters, he would first mention it casually, as a possibility, to give his men time to get used to it. For example, it became clear to him that the dogs were using up vital supplies and could no longer be maintained by the expedition. But it was also clear that Hurley had become very attached to the animals. So instead of giving a peremptory order, Shackleton briefly mentioned the problem a number of times. When he finally announced that the dogs must be put down, Hurley had been given the time to contemplate the situation and was resigned to the outcome.

* * *

Whether a team is marooned on the ice or gathered around the conference table, effective leadership requires communication. It's important to keep your team members informed about what's going on, and to speak honestly about challenges and difficulties. It's important to let them know how you plan to meet those challenges. And remember that communication can take many forms. It may not always be through a sit-down meeting in the boardroom or a mass email forwarded throughout the office. In sensitive situations, a good leader finds the right way to communicate most effectively, sometimes using a less traditional approach that won't distress its recipient but still manages to get the point across. And remember, as Shackleton did, that timing is important too – while you should opt for direct communication with your team, breaking bad news may be easier to handle if not done immediately following a stressful situation.

And make it a priority to listen as well as talk. Let your team know that you value what they have to say and will listen to constructive suggestions and ideas from all quarters. The modern workforce is more heterogeneous than ever before, and a good leader will capitalize on different perspectives from diverse backgrounds. Listening sends the message that everyone can make a contribution to solutions and successes – the very definition of a "team effort." Besides, you never know when a good idea will come your way unless you take the time to listen for it. You also may benefit from getting employee feedback that exists beyond management's usual field of vision. Whether it's fifty below zero or the heat is on, good communication will produce your team's best efforts.

GE's Jack Welch understood the importance of not only communicating, but communicating effectively (He was once quoted as saying, "Be candid with everyone"). Bill Lane, Welch's speechwriter for two decades, said in a *Directors & Boards* article that Welch realized how important communication was and he made it one of his priorities. Lane said that GE began to take off when Welch made essential changes in communications at GE. "Jack said, 'No more reports unless they're going to get people to change their behavior.' And to move people, he knew every pitch he made to whatever audience had to come from his gut." After all, communicating information via an office memo might have a different effect than speaking with each department personally. Likewise, Cisco's John Chambers saw communications as "fundamental" to revitalizing the company when it hit a rough patch in 2001, and Cisco's communications personnel is situated directly outside his office. Chambers now uses TelePresence, a live, face-to-face networking program, to connect with customers and utilizes a video blog to correspond with employees, who can post their own responses to management. And like Shackleton, Brightpoint's Robert Laikin kept his speeches to employees simple, yet effective: "There was no razzle dazzle, just three or four specific takeaways on our strategy for going forward."

The hugely successful financial news and software company, Bloomberg L.P., is renowned for a work environment that encourages communication on all levels. In its New York City offices, stairs and escalators link the many floors of the company together – there are no elevators. CEO Lex Fenwick believes that elevators hinder communication and cause people to be quiet. He'd prefer Bloomberg employees bump into one another on the stairs or escalators, where they have a better chance at striking up conversations. The same philosophy resides behind the "Link" – an open, sun-filled area that connects all parts of the company. The area is littered with numerous gathering spaces, like a group of colorful chairs and benches.

Further, there is no cafeteria but instead an open "pantry" filled with snacks and assorted beverages, where employees eat and drink amongst each other, fostering conversations one normally doesn't have when eating lunch at their desk. A furnished terrace encourages workers to socialize outside during warmer months. Many companies cannot offer such amenities, but simply having a comfortable central area where all employees can sit together and relax during a break can open the lines of communication, thus making your organization a more cohesive unit. It seems like a small detail, but Fenwick says a pantry area is powerful. "Every time I walk through the kitchen I see someone I kept meaning to talk to," he says.

And, in line with the company policy on transparency, walls in offices and conference rooms are made of glass, and employees literally work alongside each other. This lack of walls not only encourages a culture of communication, but reduces boundaries often set up between employees and their superiors. Those who work at Bloomberg also exist without official titles.

"What you have to do is break down the barriers, like titles, offices, walls," Fenwick says. "You're breaking down hierarchical barriers, and you're therefore breaking down communication vehicles. If you're the CEO, you can just walk across and talk to the salesperson and say, 'How's it going?'"

Bloomberg's belief in transparency is special because it not only opens communication among different levels of hierarchy, but encourages a culture in which *all* people work together in a more engaged and personal manner. Ideas and thoughts are transmitted easily, and a sense of unity allows employees to work together and think creatively.

Gary Erickson changed the dynamics of communication at Clif Bar when he began to notice that different departments often worked in isolation. He sensed resentment between groups and even against management – some employees were suspicious of the product's ingredients because they had never seen the bakery! His first year as sole owner, Erickson took the company on a tour of the bakery, explaining the baking process and what ingredients were used in Clif Bar's products. He created ways so that employees could see how other parts of the company worked. Weekly lunches with different departments and Thursday Morning Meetings were initiated. The entire company is present at these Thursday meetings. After breakfast is served, Erickson discusses company news or tells a story; interviews are conducted with individuals from causes Clif Bar supports. Sometimes there is a guest speaker. These meetings put Erickson face to face with his entire company, fostering trust and unity, and cutting down on gossip. Beyond meetings, managers were encouraged to walk around the office and talk to people. By tearing down walls that previously masked parts of the company, Erickson revitalized employee trust.

Meg Whitman, eBay's CEO from 1998 to March 2008, guided her company under a similar principle; instead of cornering herself in an office and pulling the strings, Whitman created a democratic organization. Her style of communication applied not only to those who work for eBay, but also for its customers. By eliminating new eBay features if customers complained or by responding directly to buyers' and sellers' emails, Whitman kept an open dialogue with the very people that created eBay's success. When eBay users first wanted to buy and sell cars on the site, Whitman listened to their request and mulled the risks involved. But, recognizing that the community of users knows what it wants, she approved the introduction of cars on eBay's site.

"It's different from traditional leadership," Whitman noted. "It's usually: What does the center want to do? It's command and control. At eBay, it's a collaborative network."

Whitman extended this dialogue with her team, acknowledging that she didn't always know everything. By listening and asking the right questions, Whitman allowed her team to come up with the right answers. Senior staff were often invited to board meetings to ensure their full participation in company business.

Good communication isn't only relevant between employees and their employers; as Bloomberg and Meg Whitman demonstrate, an effective and thriving company relies on successful communication between everyone who works within the infrastructure and even the people they often serve, the customer. By talking with one another, the creativity and innovation you have been nurturing with the previous six lessons can expand and flourish. Because Shackleton's lessons are not separate tools that exist independently from one another; the lessons work best as a cohesive system that has one objective – maximizing the chances of your team successfully reaching its goals.

EPILOGUE

Why is it so hard to find leaders of Shackleton's stature in today's world? Perhaps it has something to do with the extreme nature of his challenges. Facing extraordinary adversity can sometimes call forth extraordinary capacities.

For most of us, most of the time, the stakes are not nearly as high as they were for Shackleton every day of his mission. But that doesn't mean we can't aspire to approach the work we do with the same energy, vision and skill that Shackleton brought to his job as leader.

The lessons to be learned from his example are not just lessons for polar explorers and heroes. Shackleton's leadership is a model for anyone trying to accomplish a goal that requires team effort. They are lessons that can help you create and fulfill your leadership potential within the dynamic – and often unpredictable – business world. For managers and CEOs alike, Shackleton's example of leadership is remarkably valuable.

We may not be standing on the bow of a ship or the edge of an ice floe. But no matter where we are, we can decide that every opportunity that comes our way has the potential to be extraordinary. Because no matter what challenges a business or organization faces, we can only be successful if we know how to lead employees in a positive and efficient manner. If there are cracks at the top level of management, the entire structure can crumble. In this regard, Shackleton's story – and his example – have a great deal to teach us.

And if we apply ourselves to the task of leading with some measure of Shackleton's skill, courage and drive, then I think we can dare to look forward, as he did, to "honor and recognition in case of success."

ENDURANCE KEY DATES

August 1, 1914

The *Endurance* sets sail without Shackleton from the West India Dock to Buenos Aires, Argentina.

September 25, 1914

Shackleton leaves Liverpool for Buenos Aires, where he will meet up with the crew.

October 16, 1914

Shackleton arrives in Buenos Aires. He fires and hires several crewmen.

October 27, 1914

The *Endurance* sets sail from Buenos Aires to South Georgia.

December 5, 1914

The *Endurance* leaves South Georgia for the Antarctic.

December 24, 1914

The Ross Sea Party, who were meant to drop off supplies throughout Antarctica for the Endurance crew, sets sail on the *Aurora* from Australia toward the Antarctic.

January 18, 1915

The *Endurance* becomes stuck in the ice, one day away from reaching its designated landing spot.

February 24, 1915

Attempts to free the ship fail. The men stay on the ship, but their typical routine stops.

May 7, 1915

After reaching Antarctica, ten crewmen of the Ross Sea Party are left stranded on land after the *Aurora* is pulled into the open sea. The ship and the men onboard eventually drift into the Southern Ocean.

June 22, 1915

The crew celebrates the midwinter's day festival, which marks that winter is beginning to diminish.

October 27, 1915

The *Endurance* is crushed by moving ice and the crew abandons ship. They live on ice floes.

November 21, 1915

The crew watches the *Endurance* ship fully break apart.

October 1915 - April 1916

The crew lives on ice floes, periodically dragging their lifeboats in attempts to walk closer to land.

April 2, 1916

The *Aurora* lands in New Zealand.

April 9, 1916

A crack breaks open the ice that the men are camping on. The crew launches their lifeboats. They hope to reach whaling stations scattered on the South Shetland Islands.

April 12, 1916

Currents and winds force the crew off their track. They have no choice but to attempt to make land at the remote Elephant Island, where human contact is unlikely.

April 15, 1916

The men land on Elephant Island.

April 24, 1916

Shackleton, Worsley, Crean, McNeish, Vincent and McCarthy launch the *James Caird* lifeboat from Elephant Island for South Georgia. They hope to get help for the rest of the crew left behind.

May 10, 1916

The *James Caird* arrives safely at South Georgia.

May 10 - May 19, 1916

Shackleton, Crean and Worsley rest before beginning their march on May 19 towards Stromness whaling station, where they can seek help from its inhabitants.

May 20, 1916

The men reach Stromness whaling station.

May 23, 1916

Shackleton, Worsely and Crean board the *Southern Sky* to save the rest of the crew on Elephant Island. Ice forces them to land at Port Stanley, Falkland Islands.

June 10, 1916

After contacting various governments for help, Uruguay lends Shackleton the *Instituto de Pesca No. 1*. The ship must retreat again due to ice.

August 25, 1916

Shackleton launches the *Yelcho* ship, donated by the Chilean government, from Punta Arenas for Elephant Island.

August 30, 1916

The *Yelcho* reaches Elephant Island. All men are found alive.

Autumn 1916

Shackleton and Worsley travel aboard the *Parismina* toward New Zealand in order to save the stranded crew of the Ross Sea Party.

December 1916

Shackleton arrives in New Zealand to assist in rescue efforts for the 10 members of the Ross Sea Party stranded in Antarctica. Only seven of the 10 crewmen survived.

ACKNOWLEDGEMENTS

This book has been a long labor of love, and I have worked with some great folks to bring it to fruition. They include Ronni Gussin; Mina Samuels; Alan Goeman and eSlide; Jeanne Plant, Pamela Brooks and Alida Zamir at Taylor & Ives; Ellyn Spragins; Tricia Turnstall; Victoria Fishel; Susanne Simpson; Scott Polar Research Institute; Royal Geographical Society; Jeremy Bronson; and Matthew Amonson. Most of all, I thank my extraordinarily dedicated assistant, Korenne Haller, who has been my guiding beacon in every aspect of this project.

BIBLIOGRAPHY

Alexander, Caroline. *The Endurance: Shackleton's Legendary Antarctic Expedition.* London: Bloomsbury, 1998.

Bacon, Perry. "Obama Ratchets Up Criticism of Clinton." *Washington Post.* 28 Oct. 2007. Web. 8 Sept. 2009. <http://www.washingtonpost.com/wpdyn/content/article/2007/10/27/AR2007102701332.html>

Baker, William F., and Michael O'Malley. *Leading with Kindness: How Good People Consistently Get Superior Results.* New York: AMACON, 2008.

Bick, Julie. "Inside the Smartest Little Company in America." *Inc.* 1 Jan 2002. Web. 8 Sept. 2009. <http://www.inc.com/magazine/20020101/23798.html>

Bickel, Lennard. *Shackleton's Forgotten Men.* New York: Thunder's Mouth Press, 2000.

Birla, Madan. *FedEx Delivers: How the World's Leading Shipping Company Keeps Innovating and Outperfoming the Competition.* Hoboken: John Wiley & Sons, Inc., 2005.

Bloomberg, Michael. *Bloomberg on Bloomberg.* New York: John Wiley & Sons, Inc., 2001.

Chu, Jeff, and Kate Rockwood. "CEO Interview: Costco's Jim Sinegal." *Fast Company.* 13 Oct. 2008. Web. 8 Sept. 2009. <http://www.fastcompany.com/magazine/130/thinking-outside-the-big-box.html>

Collins, Jim, and Jerry I. Porras. *Built to Last: Successful Habits of Visionary Companies*. New York: HarperCollins, 2002.

"Csr Best Practice: Patagonia." *Article 13: The Responsible Business Experts*. Nov. 2006. Web. 7 July 2009. <http://www.article13.com/A13_ContentList.asp?strAction=GetPublication&PNID=1245>

WLIW21. "Eileen Fisher: The Wholeness Philosophy." *WLIW21: Leading with Kindness: How Good People Consistently Get Superior Results*. 7 July 2008. Web. 8 July 2009. <http://www.wliw.org/leadingwithkindness/profile/eileen-fisher/35/>

Erickson, Gary. *Raising the Bar: The Story of Clif Bar & Co.* San Francisco: Jossey-Bass, 2004.

"The Google Corporate Culture." *Netizen*. 14 June 2009. Web. 10 July 2009. <http://technbiz.blogspot.com/2009/06/google-corporate-culture.html>

Google Corporate Information. "Our Philosophy." Web. 10 July 2009. <http://www.google.com/corporate/tenthings.html>

Guth, Robert A. "Gates-Ballmer Clash Shaped Microsoft's Coming Handover." *Wall Street Journal*. 5 June 2008. Web. 10 June 2009. <http://online.wsj.com/article/SB121261241035146237.html>

Hebert, Robert. "The Second-in-Command: The Classic Profile, Making the Relationship Work and Why it So Often Fails." *The Stonewood Perspective: A Stonewood Group Inc. Bulletin*.. n.d. Web. 5 July 2009. <http://www.accesssearchpartners.com/articles/Second%20in%20Command.pdf>

Herbst, Moira. "The Costco Challenge: An Alternative to Wal-Martization?" *Working Life*. 5 July 2005. Web. 8 June 2009. <http://www.workinglife.org/wiki/The+Costco+Challenge:+An+Alternative+to+Wal-Martization%3F+(July+5,+2005)>

Hillstrom, Kevin, and Laurie Collier Hillstrom. *The Industrial Revolution in America, Volume 2*. Santa Barbara: ABC-CLIO.

Hirshberg, Gary. *Stirring it Up: How to Make Money and Save the World.* New York: Hyperion, 2008.

Huntford, Roland. *Shackleton.* New York: Carroll & Graf Publishers, 1985.

Johnson, Caitlin A. "Eileen Fisher's Unique Business Model." *CBS News Online.* 21 Jan. 2007. Web. 10 June 2009. <http://www.cbsnews.com/stories/2007/01/21/sunday/main2381285.shtml>

Kaushik, Avinash. "10 Insights From 11 Months Of Working At Google." *Occam's Razor.* 11 Feb. 2008. Web. 10 July 2009. <http://www.kaushik.net/avinash/>

Knudson, Leslie. "Perfect Package." *Business Management.* June 2007. Web. 15 June 2009. <http://www.busmanagement.com/article/Perfect-package/>

Lansing, Alfred. *Endurance: Shackleton's Incredible Voyage to the Antarctic.* New York: Carroll & Graf Publishers, 1959.

Lansing, Alfred. *Endurance: The Greatest Adventure Story Ever Told.* New York: Carroll & Graf Publishers, 1999.

Magee, David. *The John Deere Way.* Hoboken: John Wiley & Sons, Inc, 2005.

Marsh, Peter. "Difficult Furrow to Plough." *Financial Times.* 9 March 1999.

Maxwell, John C. *The 21 Irrefutable Laws of Leadership.* Nashville: Thomas Nelson, 2007.

Max, Sarah. "Seagate's Morale-athon." *Business Week.* 3 April 2006. Web. 15 June 2009. <http://www.businessweek.com/magazine/content/06_14/b3978085.htm>

McPherson, James M. "Team of Rivals: Friends of Abe." *New York Times.* 6 Nov. 2005. Web. 10 May 2009. <http://www.nytimes.com/2005/11/06/books/review/06mcpherson.html>

McSpirit, Kelly. "Sustainable Consumption: Patagonia's Buy Less, But Buy Better." *Corporate Environmental Strategy.* 5.2 (Winter 1998): 32-40.

Meyers, William. "Keeping a Gentle Grip on Power." *U.S. News & World Report.* 31 Oct. 2005. Web. 15 June 2009. <http://www.usnews.com/usnews/news/articles/051031/31whitman.htm>

Morrell, Margot, and Stephanie Capparell. *Shackleton's Way.* New York: Viking, 2001.

Murray, Matt. "Investors Like Backup, but Does Every CEO Require a Sidekick?" *Wall Street Journal.* 24 Feb. 2000. Web. *Financial Express.* 10 May 2009. <http://www.financialexpress.com/old/fe/daily/20000225/fco25082.html>

NOVA: Shackleton's Voyage of Endurance. Dir. Alan Ritsko. Videocassette. WGBH, 2002.

Nuttall, Chris. "On a Hard Drive to Create Team Spirit." *Financial Times.* 27 Jan. 2008. Web. 10 June 2009. <http://us.ft.com/ftgateway/superpage.ft?news_id=fto012720081210455076&page=2>

O'Brien, Jeffrey M. "Zappos Knows Hot to Kick it." *Fortune.* 22 Jan. 2009. Web. 26 Jan. 2009. <http://money.cnn.com/2009/01/15/news/companies/Zappos_best_companies_obrien.fortune/index.htm>

Palmer, Christopher. "Now for Sale, the Zappos Culture." *Bloomberg Businessweek.* 11 Jan. 2010: 57.

Perkins, Dennis N.T. *Leading at the Edge: Leadership Lessons from the Extraordinary Saga of Shackleton's Antarctic Expedition.* New York: Amacom, 2000.

Porr, Jeff. "'We Can Take This Hill:' CEO-talk in Hard Times: How Corporate Leaders Can Use Communications to Rebound During an Economic Downturn." *Directors & Boards.* 1 Jan. 2009. Web. *Entrepreneur.* 15 July 2009. <http://www.thefreelibrary.com/'We+can+take+this+hill:'+CEO-talk+in+hard+times:+how+corporate...-a0194976846>

Rooney, Jennifer. "Success by Design." *Sales and Marketing Management.* 1 Nov. 2005. Web. 8 Aug. 2009. *All Business.*

<http://www.allbusiness.com/marketing-advertising/4289531-1.html>

Schultz, Howard. *Pour Your Heart Into It: How Starbucks Built a Company One Cup at a Time.* New York: Hyperion, 1997.

Shackleton, Ernest. *South.* New York: Signet, 1999.

Shackleton, Ernest. *The Heart of the Antarctic.* New York: Carroll & Graf Publishers, 1999.

Shackleton: The Greatest Survival Story of All Time. Dir. Charles Sturridge. DVD. A&E Home Video, 2002.

Spector, Robert. *Amazon.com: Get Big Fast.* New York: HarperCollins, 2002.

Steinberg, Don. "Zappos Finds a Use for Twitter. Really!" *Inc.* 5 June 2008. Web. 15 June 2009. <http://www.inc.com/articles/2008/06/zappos.html>

Tyerman, Robert. "Two's Company." *Business XL.* 1 Nov. 2004. Web. 15 July 2009. *Growth Business.* <http://www.growthbusiness.co.uk/channels/growth-strategies/business-expansion/28/twos-company.thtml>

Tyler-Lewis, Kelly. *The Lost Men: The Harrowing Saga of Shackleton's Ross Sea Party.* New York: Viking, 2006.

Vise, David. A. *The Google Story: Inside the Hottest Business, Media and Technology Success of Our Time.* New York: Bantam Dell, 2008.

Vise, David A. "Tactics of 'Google Guys' Test IPO Law's Limits." *Washington Post.* 17 August 2004. Web. 8 June 2009. <http://www.washingtonpost.com/wp-dyn/articles/A6742-2004Aug16.html>

Zappos.com 2008 Culture Book, Zappos.com Inc., 2008.

NOTES

The Story

"This magnificent gift…" Huntford, 377
"[The] first crossing…" Shackleton, xix
"[W]e were taking…" Shackleton, xxvii
"splendid having Sir…" Huntford 386
"The land showed…" Shackleton, 33
"Shackleton at this…" Huntford, 415
"The disaster had…" Shackleton, 84
"I thanked the…" Shackleton, 84
"the job was…" Morrell, 133
"A man under…" Shackleton, 91
"I took it…" Shackleton, 260
"It gave one…" Shackleton, 109
"It will be…" Shackleton, 91
"All hands were…" Shackleton, 114
"a single case…" Shackleton, 117
"get into our…" Huntford, 491
"No news…Waiting…" Huntford, 479
"They were disinclined…" Shackleton, 170
"A boat journey…" Shackleton, 172-3
"I told Wild…I relied upon him…I trusted the…"
    Shackleton, 173-4, 181
"There was hope…" Shackleton, 181
"It was a…" Shackleton, 198
"Over on Elephant…" Shackleton, 205
"funny-looking" Shackleton, 227

"started as if…" Shackleton, 227
"Me – I turn…" Morrell, 190
"It might have…" Morrell, 194
"I [am] Commander…" Huntford, 641
"the most stupendous…" Shackleton, 231

Lesson One

"[Shackleton] had qualities…" Huntford, 272
"The task was…" Shackleton, 85
"invariably raises hell…" Morrell, 111
"His first thought…" Huntford, 456
"I pray God…" Huntford, 456
"I wanted to…" Morrell, 21
"[l]oneliness is the…" Shackleton, 131
"Wanted to make…" Erickson, 4
"You sell your…" Erickson, 10
"Operating from the…" Erickson, 10
"big dog eat little dog world" Erickson, 96
"I have to…" Hirshberg, 3
"It wasn't until…" Schultz, 23
"democratize the automobile…" Collins & Porras, 97

Lesson Two

"It's the old…" Morrell, 38
"energy, initiative, and resource" Shackleton, 261
"quick brain" Shackleton, 137
"unfailing optimism" Shackleton, 256
"He acts as Sir Ernest's…" Morrell, 59
"calm, cool or…" Huntford, 403
"It is largely…" Shackleton, 261
"From now on…" Collins & Porras, 169
"I'm not going…" Guth
"He could be…" Tyerman
"I'm very lucky…" Hebert, 2
"I saw hiring…" Murray

Lesson Three

"It always looked…" Morrell, 60
"their science or…" Morrell, 56
"entirely lost his…" Huntford, 212
"quite capable of…" Huntford, 214
"Notice: Men Wanted…" Morrell, 55
"The men selected…" Morrell, 56
"Many a wise…" Lansing, *Endurance: The Greatest Adventure*, 15
"perpetually chipper" O'Brien
"home"…"family" Zappos
"Create Fun and…" Zappos, 23
"We do our…" O'Brien
"In the world…" Spector, 106
"hire other great…" Spector, 107
"when you are…" Spector, 107
"improved communication, greater…" O'Brien
"We encourage our…" Steinberg
"It's not about…" Nuttall

Lesson Four

"Certainly a good…" Morrell, 89
"At present I…" Morrell, 113
"I must say…" Huntford, 425
"That was the…" Morrell, 146
"When everyone else…" Morrell, 117
"No. 1 Park Lane" Morrell, 109
"The Ritz" Shackleton, 44
"I knew how…" Shackleton, 100
"there seemed always…" Shackleton, 69
"[T]he depression occasioned…" Shackleton, 100
"soon neutralized any…" Shackleton, 109
"Meals are invariably…" Shackleton, 101
"We think a lot…" Vise, "Tactics…"
"you can be…"… "work should be …" Google Corporate Information

"Zen Oasis" Kaushik
"one of our…" "The Google Corporate Culture"
"We don't take…" Erickson, 88
"cannot build great …" "Csr Best Practice: Patagonia"
"impressive" McSpirit
"We owe that…" Chu
"From day one …" Herbst
"I think it…" Johnson
"We do it…" WLIW21, "Eileen Fisher"
"This is a…" WLIW21, "Eileen Fisher"
"You know, if…" Johnson

Lesson Five

"A man must…" Morrell, 145
"It is hard…" Shackleton, 82
"did not rage …" Morrell, 107
"Ship and stores…" Morrell, 145
"seriously worn and…" Shackleton, 147
"lips were cracked…" Shackleton, 147
"Well-settled plans …" Morrell, 167
"As always with …" Morrell, 145
"With iPod, listening…" Apple press release, 10/23/2001
"Hearing a clever …" Bick
"Looking back on…" Birla, 19
"the information about…" Fedex.com
"People are engaged…" Knudson
"It is the…" Birla, 194
"We should not…" Marsh
"I don't remember …" Magee, 64
"In trouble, danger …" Morrell, 211

Lesson Six

"Put footstep of…" Morrell, 149
"He collected with…" Morrell, 140
"a great admiration…" Huntford, 463

"be obedient to…" Huntford, 476
"I shall never…" Huntford, 476
"prairie lawyer" McPherson
"We needed the…" McPherson
"without a policy…" McPherson
"the best and…" McPherson

Lesson Seven

"[H]e would get…" Morrell, 117
"It must have…" Huntford, 455
"It was a…" Morrell, 133
"I decided not…" Shackleton, 159
"Be candid with…" www.brainyquote.com
"Jack said, 'No…" Porr
"There was no…" Porr
"Every time I…" Rooney
"What you have…" Rooncy
"It's different than…" Meyers
"ducking the issue" Bacon

For information on Arthur Ainsberg's leadership lecture, *Shackleton: Leadership Lessons from Antarctica*, please contact Arthur through his website:

www.AinsbergOnLeadership.com

CPSIA information can be obtained at www.ICGtesting.com
Printed in the USA
LVOW122134030412
276062LV00001B/26/P